Ecology of Birds

ECOLOGY OF BIRDS

AN AUSTRALIAN PERSPECTIVE

by

Hugh A. Ford

Published by

Surrey Beatty & Sons Pty Limited

ISBN 0 949324 19 1
National Library Canberra
Australia

Published June, 1989

PRINTED AND PUBLISHED IN AUSTRALIA BY
SURREY BEATTY & SONS PTY LIMITED
43 Rickard Road, Chipping Norton, NSW 2170

AUSTRALIAN ECOLOGY SERIES

General Editor: Harold Heatwole

Surrey Beatty & Sons (current Ecology Series publisher)

Ecology of Reptiles by Harold Heatwole and Janet Taylor

Ecology of Birds: an Australian Perspective by Hugh A. Ford

Other Titles in this series:

(University of Queensland Press: former publisher)

Insect Ecology by E. G. Matthews and R. L. Kitching

Ecology of Marine Parasites by Klaus Rohde

Seashore Ecology by Thomas Carefoot and Rodney Simpson

Ecology of Mangroves by Patricia Hutchings and Peter Saenger

Dedicated to my Parents and Parents-in-Law

Contents

Illustrations and Tables

FIGURES (Black and White)

PLATES (Colour)

TABLES

Foreword

Ecology is the science involved with the interactions of organisms and their physical and biotic environments. This field has always been a source of fascination to professional biologists, naturalists and conservationists. In recent years, as the human population has progressively increased, environmental problems have become of vital interest and importance to the public as well. It has now become imperative that ecological principles and the ecology of specific regions and taxa be understood by a wide variety of people. The present series is designed to help fill this need.

It is felt that the volumes in this series will serve as a source of information for university students, teachers and the interested public who require a basic factual knowledge to broaden their understanding of ecology, and for those conservationists, agriculturists, foresters, wild life officers, politicians, engineers, etc., who may need to apply ecological principles in solving specific environmental problems. In addition, it is hoped that the series will be a valuable reference work and source of stimulation for professional ecologists, botanists and zoologists. The writing is at a level that will neither encumber the layman with unnecessary jargon nor be too elementary to be of interest to the professional ecologist.

The study of ecology can be approached on various levels. For example, one can emphasize the biotic community and analyse the kinds and numbers of organisms living together in a particular habitat, the way they are organized in space and time and the interactions they have with each other. This type of ecology is known as synecology.

Another way of studying ecology is by systems analysis. In this method, the biotic community and the physical environment which together make up what is known as an ecosystem, are looked upon as a functioning unit. In such an approach the main emphasis is on the cycling of energy, minerals or organic materials within the ecosystem and the factors influencing these processes, rather than specifically upon the organisms themselves. Often mathematical or theoretical models are constructed and tested, frequently with the aid of computers.

Both of the above approaches are synthetic; they take an overview of entire communities or systems and do not emphasize individual species. By contrast the following two approaches, collectively known as autecology, are concerned mainly with particular species.

The population approach, often called demography, is concerned with: (1) fluctuation in the abundance and distribution of individuals of a given species in an area; (2) the contributing phenomena such as birth and death rates, immigration, emigration, longevity and survival; and (3) the influence of the physical environment and of other species on these characteristics. Of major interest are mechanisms regulating population density and factors influencing population stability.

The final approach to ecology is one primarily concerned with the effect of the environment on the individuals of a species, that is, how they are affected by temperature, moisture, light or other external factors. This approach is known variously as environmental physiology or physiological ecology. The keynote is adaptation to specific environments.

All of the above approaches will be employed with varying emphasis in the volumes of this series.

Certain topics, such as ecology of grasslands, ecology of forests and woodlands, or ecology of deserts lend themselves to a community approach; grassland, forest and desert *are* types of communities and if studied as an entity must be approached on the community or ecosystem level. On the other hand, where specific taxa such as reptiles, birds or mammals are treated, the autecological approach is more often used. The particular aspect emphasized varies from group to group, depending on the information available.

Regardless of emphasis, in each book of this series the available information in a particular field is reviewed critically and summarized, so that the reader is brought abreast of current knowledge and developments. Recent trends are indicated and the foundations for future developments are prepared by highlighting conspicuous gaps in knowledge and pointing out what appear to be fruitful avenues for research.

Harold Heatwole

PREFACE

Birds are watched and enjoyed by increasing numbers of amateur naturalists around the world. They are popular because they are colourful, active and diurnal, and, like us, communicate by sight and sound. More serious studies on birds have contributed to the development of ecological theory, especially in the fields of population and community ecology. Birds have also been favourite subjects for the study of animal behaviour, particularly foraging, territorial and reproductive behaviour. In addition, ornithologists have been at the forefront of the conservation movement.

Most of the early studies of avian ecology and many of the significant recent ones have been carried out in Europe or North America. Consequently, most of the framework for avian ecological theory has emerged from these two continents. However, the environments of tropical regions and the southern continents differ markedly from those to the north, so that ecologists are now questioning how well the generalizations derived from northern temperate parts of the world stand up to scrutiny elsewhere.

The last two decades have seen rapid developments in the knowledge of the ecology of birds in Australia. Australian birds differ from those of other continents; indeed they are more unique than was even believed five years ago. Despite this, many of the patterns observed in Australian bird assemblages are familiar and predictable. A few are not: such as the high incidence of nectar-feeding birds found in temperate regions and the abundance of co-operatively breeding species. In this book I shall introduce a series of ecological topics from their current theoretical basis and then use such information as is available from the Australian region to examine these theories. Although I emphasize studies in Australia, I also use, where appropriate, work from New Zealand, New Guinea and elsewhere in the Pacific region. I hope to investigate the extent to which the theories explain the facts, or, if not, how the theories need to be modified. Often the data are too scanty so I stress areas in which more information is needed and discuss how one could collect it.

I also discuss the impact of humans on the birds of the region and consider how work elsewhere plus the peculiarities of Australian environments should influence the ways in which bird populations should be managed.

The ideas and information presented in this book result from the written work of and extensive discussions with many others. In particular I should like to thank the following for critically reading one or more chapters: David McFarland, David Paton, Stephen Marchant, Ed Wyndham, Stuart Pimm, Ray Nias, Jiro Kikkawa, Kim Lowe, Joost Brouwer, David Baker-Gabb, Claire Appleby and Harry Recher. I am grateful to Harold Heatwole for his advice throughout the writing of the book. My wife, Gill, has been a constant source of encouragement, especially in the final stages. I also thank Sandra Higgins and Viola Watt for typing numerous drafts of the manuscript, Lynda Bridges for drawing the figures and Bob Hardie, Col Mulquiney, Bob Howe, Rodney Simpson, Harold Heatwole, the Otago Peninsula Trust and the Australian National Photographic Index for providing the photographs.

Frontispiece: The inconspicuous Speckled Warbler (*Sericornis sagittatus*) epitomizes our knowledge of Australia's birds. Common in suitable habitat, much of its ecology and behaviour remains a mystery. It could be polygamous, or co-operative, as two brief papers suggest, or even both. It also eats fruit, but does it disperse or digest the seeds?

1. The Australian Environment and the Origin of Birds

Before one can understand the ecology of Australian birds one needs to know what sort of environment they occupy. One also needs to know what their environment was like in the past. When did birds arrive in Australia, and from where? How have they changed since they came here and why? In what ways are Australian birds unique?

The Environment: Past and Present

One third of Australia is now tropical, the rest largely sub-tropical with a temperate southern fringe. It is usually perceived as a hot, sunny and dry country. Indeed it is true that over 50% of the continent can be termed desert, either as sand-dunes covered in prickly grass or endless stony plains of dismal saltbush. It has not always been like this. In the early Tertiary period, around 50 million years ago, Australia was some 20° south of its present position. It was part of the great southern continent of Gondwana and was attached to Antarctica and hence to South America (Barlow 1981). Evidence from pollen analysis and other fossils indicates that the predominant vegetation was rainforest. Trees such as southern beech (*Nothofagus*) and podocarps were abundant as were ancestors of the modern hoop-pine, *Araucaria*. Certainly many plants that were less prolific pollen producers would have been common too, and we should not visualize Australia as being clothed in forest like that found in the few remaining mainland relicts of *Nothofagus* forests. The climate would have been wetter and the soils richer than today. The latitude of Australia then suggests that it would have been cooler than today but worldwide temperatures were apparently higher. Perhaps all we can say is that solar radiation would have been weaker, especially in winter. Maybe it was more seasonal than it is now.

Between 40 and 50 million years ago Australia separated from Antarctica and started drifting northwards. It moved into the tropics and some 10 million years ago approached southeast Asia. During the mid-Tertiary the climate became drier and a sclerophyllous (tough-leaved) flora emerged from the rainforests, probably originating on infertile ridges and outcrops from which nitrogen and phosphorus had been leached. Rainforests remained on the coasts but the centre of the continent became progressively drier. The last 2–3 million years (Pleistocene and Recent) have seen wide fluctuations between arid and moist conditions. The ice ages of the temperate regions were dry periods for most of Australia, the interglacials were pluvials with real lakes in central Australia and nearly continuous rainforest (Barlow 1981; Lange 1982; Singh 1982). The world is currently in, and is perhaps drawing to the end of, a favourable period. It appears to have been moister 8,000 years ago (Singh 1982) but was much drier 20,000 years ago.

The Origin of Australian Birds

Generally, birds make very poor fossils, as they have small and light bones, which are not very distinctive. Therefore we know very little about the kinds of birds that inhabited Australia until recently (Rich and Baird 1986). We have to infer, from worldwide distributions, which bird families might have been present when Australia was part of Gondwana, and which arrived much more recently from Asia. In the last few years biochemical evidence has been used to piece together the relationships of Australian passerines with those of other continents. This has also allowed tentative dates to be placed on the main radiations of passerine birds.

The oldest Australian bird fossil is a feather from Victoria, about 120 million years old (Rich and Van Tets 1982). This is only about 40 million years younger than *Archaeopteryx*, the earliest known bird fossil in the world, and indicates that birds have had a long history on the Australian continent. The first identifiable fossil Australian birds were giant penguins (*Pachydyptes*), dating from about 45 million years ago. The present distributions of penguins is almost entirely southern, with one equatorial species, which strongly indicates a Gondwanan origin. By the late Miocene (about 20 million years ago) central Australia had extensive wetlands as indicated by fossil cormorants, pelicans, ducks, geese, gulls, rails, cranes and flamingoes. The flamingoes are especially interesting as Australia is the only continent, except Antarctica, where flamingoes are not found today. It has been suggested that the flamingoes' nearest relatives are Banded Stilts (*Cladorhynchus leucocephalus*[1]), which are endemic to Australia (Olson and Feduccia 1980). Presumably flamingoes, and many other waterbirds, were lost during the dry Pleistocene periods.

The first landbird fossils from Australia are stone-curlews, owlet-nightjars and mihirungs. Mihirungs were giant ratites, the largest of which, *Dromornis*, weighed about 450 kg (Rich and van Tets 1982). Ratites are clearly another Gondwanan group with modern representatives in South America (rheas), Africa (ostriches), Australia (emus, cassowaries) and New Zealand (kiwis), plus recently extinct species from Madagascar (elephant birds) and New Zealand (moas).

The Pleistocene fossil record is more extensive but is mostly of modern species, plus some, such as a giant megapode and an eagle, that have subsequently become extinct. According to their modern distribution and diversity in Australia and other southern continents several non-passerine groups could well have had a southern origin (Rich 1975). These include the parrots, and the megapodes — the hen-like mound-builders, such as the Malleefowl *(Leipoa ocellata)* and Brush-turkey (*Alectura lathami*). Other possibilities are pigeons and bronze-cuckoos (*Chrysococcyx* — Marchant 1972).

The passerine order of birds contains over half of the known species and is usually considered to be the most modern. Passerines are popularly called perching birds or songbirds and include the lyrebirds, flycatchers, finches, thrushes, warblers, swallows and almost all the smaller landbirds. Before the general acceptance of the theory of continental drift it was assumed that Australia received most of its landbirds from Asia, by island-hopping (Mayr 1944). Several species like the swallows, pipit and silvereyes are so similar to widespread northern forms that they are obviously recent arrivals (Table 1). Others, like the honeyeaters, are obviously based in Australia and have had a long history there. The most enigmatic birds, though, are the ones that the early naturalists called treecreepers, thrushes, robins, warblers, wrens and flycatchers. These resembled in shape and behaviour birds of the same names in Europe, but were they really closely related? How long had they been in Australia?

Did any passerines reach Australia before it was isolated from Gondwana or have they all arrived recently from Asia or could they have crossed the formidable sea barriers during the 20–30 million year period when it was remote from the other continents? Unfortunately classical taxonomic methods cannot be used to

[1]Scientific names of birds are given at the first mention of a species in each chapter.

Table 1. Families of passerines that have arrived recently in Australia from Asia, and now breed (excluding human introductions).

Pittidae (Pittas — 3 species)
Turdidae (Thrushes — 2)
Sturnidae (Starlings — 1)
Hirundinidae (Swallows — 4)
Cisticolidae (Cisticolas — 2)
Zosteropidae (Silvereyes — 3)
Sylviidae (Warblers — 6)
Alaudidae (Larks — 1)
Nectariniidae (Sunbirds, Flowerpeckers — 2)
Ploceidae (Grassfinches, Pipits — 19)

indicate the degree of relationship between Australian passerines and those of other continents for two reasons. First, many resemblances between species are the result of convergence due to similar selective forces brought about by a similar way of life. Second, changes may take place very rapidly when animals reach new environments. This is illustrated best by the finches of the Galapagos (Geospizinae — Lack 1983) and Hawaiian islands (Drepanididae — Raikow 1976). New techniques were needed and these now have been provided by Sibley and his colleagues. They have looked at biochemical similarities, first using proteins and later using DNA. Their findings, based on comparisons of the DNA of over 1,000 species, have revolutionized our thinking about the position of Australia in bird evolution.

Sibley and Ahlquist (1985) showed that almost all Australian passerines are more closely related to each other than to species of similar appearance elsewhere. They represent a massive adaptive radiation from a stock that colonized some 50 million years ago. The relationships of the main groups are shown in Figure 1. It was a major surprise to find that the Australian treecreepers were related to the lyrebirds and bowerbirds and not to tree-creeping birds from other continents. Likewise the Australian wrens and warblers are related to the honeyeaters and perhaps most amazingly Australian robins, thrushes, flycatchers and many other native birds are related to crows!

This new classification is not only important to taxonomists, it is exciting to ecologists too. It shows that Australia provides an almost totally independent scene in which to test the ecological hypotheses that have been generated in northern temperate regions. Questions such as: how are bird communities constructed? how are populations regulated? how do co-existing species affect each others' foraging ecology? and how do isolated populations adapt to new environments? have been answered principally in studies carried out in the north-temperate environments where the ideas originally were generated. Now these questions can be asked all over again in a different theatre with different actors.

Sibley and Ahlquist provided a framework for the evolution of Australian passerines and an approximate time-scale, although they have not revealed where the first passerines originated.

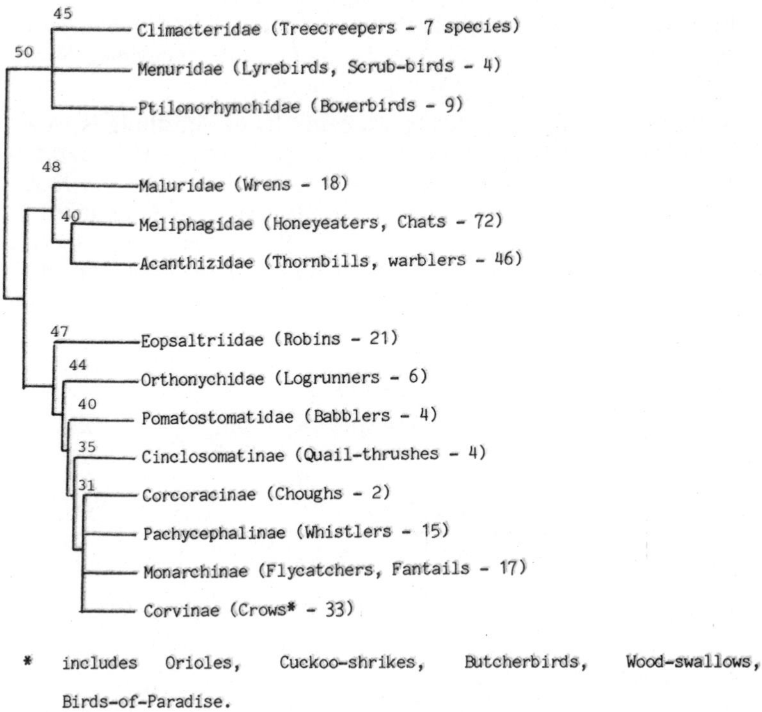

Fig. 1. Relationships, numbers of Australian breeding species and approximate dates of divergence (millions of years ago) of Australian passerines. Adapted from Sibley and Ahlquist (1985).

Divergence 50 million years ago suggests a southern origin, but Australian passerines are even more distantly related to South American passerines than they are to Asian ones. In fact New Zealand has at least one species, the Rifleman (*Acanthisitta chloris*) whose closest relatives are South American (Sibley *et al.* 1982), so Australia probably had similar birds at one time but has since lost them. The ancestors of the lyrebirds and honeyeaters must have reached Australia, across sizeable water gaps, from southeast Asia or Africa. The movements of the southeast Asian plates are not yet fully understood so perhaps there were large islands between Australia and Asia, across which birds could have travelled (Audley-Charles 1983). There would also have been islands between Africa and Australia, which would have been nearer then than they are today. Also both would have been well in the westerly belt of winds, which perhaps intensified as the southern ocean opened. Birds could have been carried from Africa to Australia, just like North American birds are sometimes blown across the Atlantic to northwestern Europe.

Wherever the early passerines came from one thing is clear: Australia has not been a mere recipient of strays. It generated many new families of birds, some of which crossed the seas back to Asia and Africa. Although the family Corvidae originated in Australia the true crows (*Corvus*) evolved elsewhere, before returning to Australia. The orioles (*Oriolus*) too have had an early and a later history in Australia. Colonization between Australia and Asia has been a two way process.

The Present Day Australian Avifauna

Today Australia has 575 breeding species, about 80% of which are landbirds and just over half are passerines (Table 2). Another 63

Table 2. The modern Australian avifauna.

Non passerines:	breeding — 275 species non-breeding — 129
Passerines:	breeding — 300 non-breeding — 8
Landbirds:	breeding — 454 non-breeding migrants — 5 vagrants — 10
Waterbirds:	breeding — 81 non-breeding migrants — 34 vagrants — 18
Seabirds:	breeding — 40 non-breeding migrants — 24 vagrants — 46

species occur as migrants; these are chiefly seabirds or waders, and another 74 should be regarded as vagrants: rare or accidental visitors. To put this in perspective there are rather more breeding species than in Europe, but slightly fewer than in North America.

New species are still being described, e.g. the Eungella Honeyeater (*Meliphaga hindwoodi* — Longmore and Boles 1983) and the Rusty-tailed Fly-eater (*Gerygone ruficauda* — Ford and Johnstone 1983) from Queensland. However, it seems likely that knowledge of birds is more complete than for any other Australian taxon.

Australian Vegetation

The main vegetation zones in Australia are arranged concentrically around the arid centre (Figure 2). Although desert, with

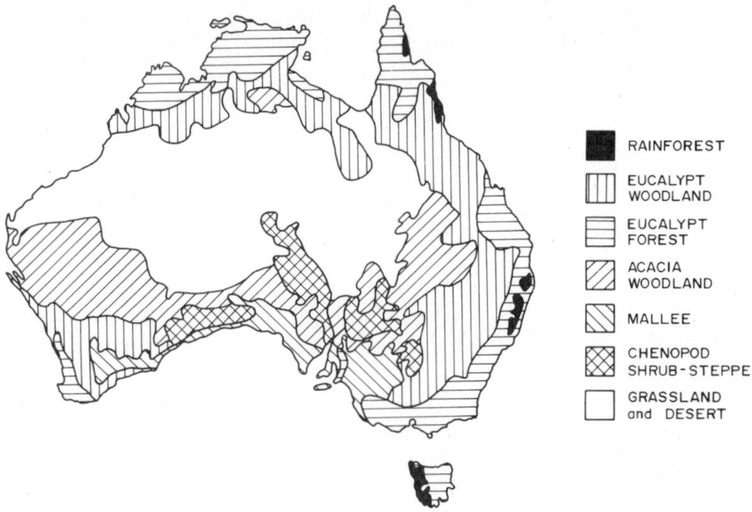

Fig. 2. Distribution of the major vegetation zones in Australia (modified from Bridgewater 1987).

below 125 mm average rainfall, occupies an extensive area, it is largely covered in vegetation and is by no means birdless. The sandy deserts are clothed in grasses, particularly spinifex (*Triodia* — Plate 1). Spinifex pigeons (*Petrophassa plumifera*), the Spinifexbird (*Eremiornis carteri*) and grasswrens (*Amytornis* — Plate 2) are characteristic of this habitat. Shrub-steppe of saltbush (*Atriplex* — Plate 3) and other succulent chenopods (Plate 4) is found in the stony deserts. Typical birds there are chats (*Ephthianura* — Plate 5) and pipits (*Anthus*).

Large areas of Australia are covered in semi-arid woodlands, usually dominated by *Acacia* or *Eucalyptus* (Plates 6, 7 and 8), but occasionally with *Casuarina*, *Callitris* (native pine) or other trees. Particularly distinct semi-arid woodlands include Mallee and Mulga. Mallee consists of many-stemmed eucalypts growing from a subterranean lignotuber (Plate 8). It has a diverse bird community, including the Malleefowl (*Leipoa ocellata* — Plate 9), ringneck parrots (*Barnardius*) and the Gilbert's Whistler (*Pachycephala inornata*). Mulga (*Acacia aneura*) is inhabited by a wide range of parrots and passerine birds such as the Redthroat (*Sericornis brunneus*) and Cinnamon Quail-thrush (*Cinclosoma cinnamomeum*). Eucalypt woodlands (Plates 10 and 11) are found between the 300 and 1,000 mm isohyets in northern Australia and between the 300 mm and 500 mm isohyets in southern Australia. They usually have a sparse understorey, but a well-developed grass layer. Many common birds such as the Rufous Whistler (*Pachycephala rufiventris*), Jacky Winter (*Microeca leucophaea*), Brown Treecreeper (*Climacteris picumnus*) and the kookaburras (*Dacelo* — Plate 12) reach their greatest density here.

True forests are restricted to within 200 km of the coast in most of Australia. These are usually dominated by eucalypts (Plates 13 and 14) with only those in the wettest places and on good soils becoming rainforests (Plates 16 and 18). Characteristic birds of eucalypt forest include the Crimson Rosella (*Platycercus elegans* — Plate 15), lyrebirds (*Menura* — Plate 17) and the Scarlet Robin (*Petroica multicolor*). Rainforests are the home of fruit-doves (*Ptilinopus*), fig-parrots (*Opopsitta*), riflebirds (*Ptiloris*), pittas (*Pitta* — Plate 19) and most of the bower-birds (Ptilinorhynchidae). Some 65 species of birds are restricted to rainforest (Frith 1976) but many others span both rainforests and eucalypt forests. Heathlands (Plate 20) are found in a variety of places: above the treeline in the Snowy Mountains and Tasmania, on deep sands in mallee country, in swampy coastal areas and on exposed seacliffs. They have very diverse and often showy floras and are most notable from an ornithological viewpoint for their diversity and density of honeyeaters (Meliphagidae — Plate 21). Mangroves (Plate 22) fringe much of the northern coastline and occur patchily in the south. They have characteristic species of heron (Plate 23), kingfisher, honeyeater and warbler (*Butorides striatus, Halcyon chloris, Lichenostomus fasciogularis*, *Gerygone laevigaster* respectively).

Perhaps surprisingly for such a dry continent, Australia has some spectacular wetlands (Plate 24). These are almost all seasonal, and often ephemeral. The northern wetlands, which shrink in the dry season, are the home of Jacanas (*Irediparra gallinacea* — Plate 25), Magpie Geese (*Anseranas semipalmata*),

Jabirus (*Xenorhynchus asiaticus*), and many water-fowl and large wading-birds (Plate 26). The Murray-Darling basin is the other major wetland, with important breeding grounds for ducks and ibis. In wet years, such as 1973-74 and 1984, the inland water-courses flood and the salt lakes fill. At such times Lake Eyre is colonized by thousands of Pelicans (*Pelecanus conspicillatus*), cormorants, gulls and terns.

Two types of coastal habitats are significant. These are estuaries and beaches for migrant waders (Plate 27), and offshore islands for breeding seabirds. Australia is host for between one and two million waders from the Northern Hemisphere in the southern summer. These are concentrated in three chief regions, Eighty-Mile Beach, Port Hedland and Roebuck Bay in north-western Australia; the Gulf of Carpentaria; and the Coorong and neighbouring Spencer and St. Vincent Gulfs in South Australia (see Figure 37). The first two only became recognized in the 1980s (Lane 1987). The country's main seabird colonies are found in the Bass Strait and on islands of the Great Barrier Reef. Phillip Island, 60 km southeast of Melbourne has the most famous Little Penguin colony (*Eudyptula minor*), though others are larger. The Shy Albatross (*Diomedea cauta*) breeds on three islands off Tasmania, whereas perhaps 8 million pairs of Short-tailed Shearwaters or Muttonbirds (*Puffinus tenuirostris*) nest in the Tasmanian region. The Cape Barren Goose (*Cereopsis novae-hollandiae* (Plate 29) breeds on islands off southern Australia. Other southern breeding seabirds are storm-petrels, prions, diving-petrels, gannets, gulls and terns. Raine Island, off Cape York Peninsula, holds the most diverse collection of tropical sea-birds, with three species of boobies (*Sula sula, S. dactylatra* and *S. leucogaster* — Plate 30), Wedge-tailed Shearwaters (*P. pacificus*), frigatebirds (*Fregata*), Red-tailed Tropicbirds (*Phaeton rubricauda*), and three species of terns (*Sterna, Anous*).

Australia makes up a major part of the Australian zoogeo-graphical region, which includes New Zealand, New Guinea and neighbouring islands and the islands of the southwestern Pacific. It has been split into several faunal subregions, based on the similarities or differences of birds or other animals within or between particular areas. The most comprehensive analysis of the avifauna is that of Kikkawa and Pearse (1969). They plotted distributions of 464 species of landbirds from 121 sites and divided Australia into 11 areas (Figure 3). The major subdivision was east from central plus west. Tasmania was also quite distinct from the mainland. Their map was remarkably similar to

Fig. 3. Faunal subregions of Australia (modified from Kikkawa and Pearse 1969). The Kosciuskan and Tasmanian subregions are often combined into the Bassian subregion.

those derived on a more intuitive basis; indeed most of the regions have well-established names (see Figure 3). Kikkawa and Pearse, however, rejected the name Bassian for the southeastern region. The Torresian and Kosciuskan are the richest in species, the Tasmanian subregion the most depauperate.

New Guinea is part of the Australian continental plate which has pushed up recently and has been attached to Australia for much of its history. Being tropical, mostly covered in rainforest and close to the Oriental Region it has a rich avifauna. Although it has only 10% of the land area of Australia it has nearly as many species (Beehler *et al.* 1986). It is the centre of distribution of the birds of paradise, bowerbirds, cassowaries, megapodes and crowned pigeons. It is also rich in pigeons, parrots, cuckoos, kingfishers, honeyeaters and Australian warblers and flycatchers. A single species each of hornbill and true shrike (Laniidae) also breed in New Guinea.

New Zealand has been isolated from other land masses for even longer than Australia. It has at times been smaller and extensively glaciated. It consequently has a depauperate avifauna, as do the other groups of Pacific Islands. There were about 42 species of breeding landbirds at European settlement, including the endemic families of New Zealand wrens (Acanthisittidae), thrushes (Turnagridae) and wattlebirds (Callaeidae). New Zealand is very rich in seabirds.

2. Distribution

Distribution is one of the most basic ecological properties of a species. Indeed Andrewartha and Birch (1954) suggested that animal ecology could be defined as the study of the distribution and abundance of populations. A wide variety of factors limits the distribution of species and if one realizes that distribution is simply the area in which a species has an abundance of greater than zero, it is not surprising that most of these factors also play a part in limiting abundance.

Many species do not realize their potential distribution because they are unable to cross barriers to reach suitable areas (Krebs 1972). For birds such barriers are often extensive stretches of sea, or high mountain ranges or unsuitable habitat. Where there are no major barriers the main limits to distribution are caused by interactions with other species, or by climate (Krebs 1972). In other cases species are limited simply because they choose particular habitats and avoid others.

Interactions with other species take a variety of forms. Perhaps most obvious is the presence of suitable plant food or animal prey. Crossbills (*Loxia*) are specialized finches that feed on the seeds of conifers in Europe and North America. Their distribution is closely tied to extensive areas of spruce or pine forest (Newton 1972). However, birds are frequently generalized in their diet and often use different foods in different parts of their range. The Northern Pied Oystercatcher (*Haematopus ostralegus*) was originally restricted to coastal parts of Britain where molluscs such as cockles (*Cardium edule*) and mussels (*Mytilus edulis*) were abundant. This century it has spread along river valleys and into grassland where it feeds mostly on earthworms (Heppleston 1972). Predators also can limit a species' range. Many species of seabirds nest on off-shore islands or steep cliffs where they can avoid mammalian predators. Introduced cats and rats have devastated many seabird colonies.

Sometimes a pair of closely related and ecologically similar species have abutting or marginally overlapping ranges. In such cases it is usually inferred that they exclude each other competitively. More interesting are the cases where two species overlap and diverge so that they differ more where they are sympatric, or conversely expand their use of resources where potential competitors are absent. Darwin's finches of the Galapagos Islands provide many examples (Lack 1983; Grant 1986). The Large Cactus Finch (*Geospiza conirostris*) occurs on islands where the Small Cactus Finch (*G. scandens*) and Medium Ground Finch (*G. fortis*) are absent. It is intermediate between these two species in bill shape and diet. On an island where the Large Ground Finch (*G. magnirostris*) is also absent *G. conirostris* has a larger beak and takes hard seeds.

Disease organisms may also limit bird distributions, though this is difficult to determine. Most of Hawaii's native birds are not now found at lower elevations than 1,000 m as avian malaria is spread there by mosquitos (Warner 1968).

Usually, though, it is hard to relate the distribution of a bird species to one or a few species of prey, predators or competitors. More often a species is found in one or a few biomes (Pitelka 1941) or habitats, for instance the northern coniferous forests have a characteristic avifauna.

Birds are perhaps limited directly by climate only under extreme conditions, such as excessive heat, aridity or cold. More often weather acts via food supply. For instance dry climates may reduce or prevent production of flowers, fruits or seeds, and snow may cover up terrestrial food. Water shortage may severely limit birds in dry areas and the species that inhabit such places show various adaptations that conserve water (Serventy 1971).

Dispersal to Australia

As mentioned in the previous chapter, Australia has been distant from the other continents for much of the last 50 million years. Many groups have not reached Australia because they were unable or unwilling to cross water gaps. Alternatively they may have been kept out by competition from birds already present. The woodpeckers are an excellent example of a group missing from Australia. They reach Indonesia but not New Guinea, despite the fact that they would probably survive there. Interestingly, woodpeckers are widespread through the West Indies having crossed water gaps far larger than those between

Indonesia and New Guinea (Lack 1976). Several bird species have, however, made the crossing in the last few million years (see Table 1), and many are very successful in Australia, such as the Silvereye (*Zosterops lateralis*). Some species have strangely missed Australia but colonized more isolated regions. The Grey-headed Blackbird (*Turdus poliocephalus*) occurs, or occurred, on Norfolk and Lord Howe Islands (Schodde *et al*. 1983). It is possible that mainland Australia was unsuitable for the species, or that competitors prevented its colonization. However, the ecologically similar Blackbird (*T. merula*) has successfully invaded natural habitats in Australia.

A total of 15 species have successfully colonized Australia after being introduced by Europeans. Many other attempts to introduce species have failed and most of the successful species have stayed within greatly altered habitats such as gardens, cities or farmland. The Goldfinch (*Carduelis carduelis*) for instance, eats mostly seeds of introduced weeds (Middleton 1970a). Northern temperate birds occasionally occur in Australia, such as the Grey Wagtail (*Motacilla cinerea*), which was found at virtually the only site in South Australia that resembles its Eurasian habitat of a mountain stream (Sara 1982). It is clearly a big step from reaching another continent to colonizing it.

Perhaps the only successful natural colonizer of Australia in historical times has been the Cattle Egret (*Ardeola ibis*), but even in this case Europeans have modified the habitat and one attempted introduction may have succeeded (Blakers *et al*. 1984). Several Australian birds have reached New Zealand very recently, such as the Welcome Swallow (*Hirundo neoxena*) and Sacred Kingfisher (*Halcyon sancta*). Compared with Australia, New Zealand has a depauperate avifauna, partly the legacy of recent glaciation and partly because species were exterminated by the Maoris and European settlers. Many introductions by Europeans have been successful.

Distribution within Australia

It is fortunate that the basic pattern of distribution of birds in Australia is now well-known. This is because some ten years ago the Royal Australasian Ornithologists' Union launched an ambitious project to map each species on a grid of 1° squares covering the whole country. Over five years some 3,000 volunteers contributed 2.7 million records on 716 species. The resulting atlas (Blakers *et al*. 1984) displays both breeding and total ranges for all of these.

To say that the distribution of a species is known does not mean that it is understood. Even before the Bird Atlas was prepared, Keast (1961) extensively explained bird distribution in Australia in terms of present or past barriers and habitats. However, there is still abundant scope for interpreting distributions and following this up with studies of the species' ecology. Two factors make interpretation of bird distributions especially difficult. First, birds are highly mobile and likely to occur well outside their "normal" range. Seabirds are sometimes found well inland and desert birds may occur on the coast. Second, distribution on a scale of 1° may be far too coarse, as birds, and most animals, respond to features on a much more local scale.

At one extreme some species cover the whole country (e.g., Brown Falcon *Falco berigera*, Willie Wagtail *Rhipidura leucophrys*, Black-faced Cuckoo-shrike *Coracina novaehollandiae*) whereas others have tiny ranges. The Noisy Scrub-bird (*Atrichornis clamosus*) occupies only a few km^2 of coastal heathy forest in southwestern Australia and the Eungella Honeyeater (*Meliphaga hindwoodi*) inhabits rainforests in a single mountain range perhaps 50 km long in eastern Queensland (Figure 4).

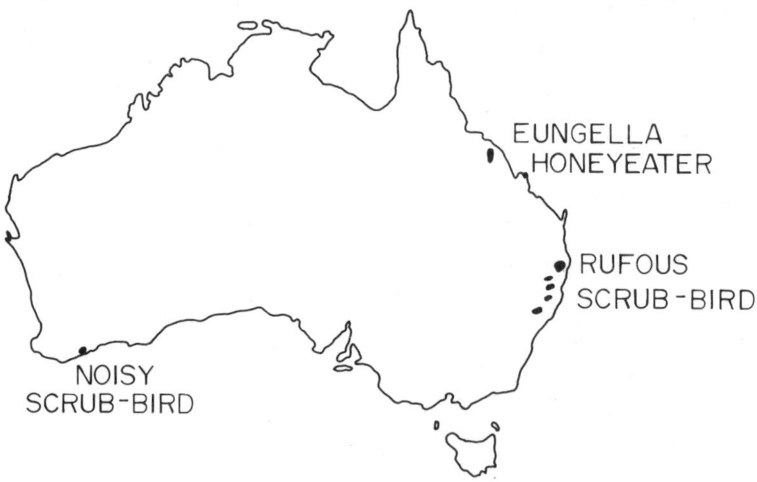

Fig. 4. Relict species. Distribution of the Eungella Honeyeater and Noisy and Rufous Scrub-birds.

The range of the Noisy Scrub-bird has been reduced in historic times (Smith 1977), but there is no reason to believe that the range of the Eungella Honeyeater has changed markedly in recent times. A few species are not only wide-ranging in Australia, they also occur on other continents. For example, the Peregrine Falcon (*Falco peregrinus*) is found in all continents except Antarctica, and the Osprey (*Pandion haliaetus*) breeds in Eurasia, North America and North Africa. A few species are widespread in mainland Australia yet absent from Tasmania, e.g., Crested Pigeon (*Ocyphaps lophotes*), White-winged Triller (*Lalage sueurii*) and Rufous Whistler (*Pachycephala rufiventris*). In contrast Tasmania has 11 endemic species, including the Yellow Wattlebird (*Anthochaera paradoxa*) and Forty-spotted Pardalote (*Pardalotus quadragintus*). The flightless Tasmanian Native-hen (*Gallinula mortierii*), previously occurred on the mainland (Baird 1984) but was eliminated there, possibly by an extremely dry period about 10,000 years ago.

Both Bass and Torres Straits constitute barriers to birds, though several species cross them annually. Orange-bellied Parrots (*Neophema chrysogaster*) and Swift Parrots (*Lathamus discolor*) breed in Tasmania, yet winter on the mainland. Rainbow Bee-eaters (*Merops ornatus*) and several species of flycatchers spend the winter in New Guinea. The most important barriers to dispersal in Australia though are the deserts. Numerous species occur as isolated populations in southwestern and eastern Australia, separated by the Nullarbor Plain and Great Victoria Desert (e.g., Scarlet Robin *Petroica multicolor* and Regent Parrot *Polytelis anthopeplus*) (Figure 5). Others are split by the dry

Fig. 5. Species with populations isolated by dry barriers. Distribution of Regent Parrot and Satin Bowerbird.

country to the south of the Gulf of Carpentaria, these include the Helmeted Friarbird (*Philemon buceroides*) and Yellow Oriole (*Oriolus flavocinctus*). The barriers need not be particularly dry ones, the Pale-Yellow Robin (*Tregellasia capito*) and Satin Bowerbird (*Ptilinorhynchus violaceus*) occur in rainforest in the Atherton region of northeast Queensland but are then absent until one reaches southeast Queensland, 1,000 km further south (Figure 5). This is despite the fact that the intervening country is mostly eucalypt forest with several patches of rainforest.

A few species have highly disjunct populations. The Crested Shrike-tit (*Falcunculus frontatus*) probably occurs as five isolates, whereas the Yellow Chat (*Ephthianura crocea*) inhabits coastal rivers in northern Australia plus some inland flood-plains (Figure 6).

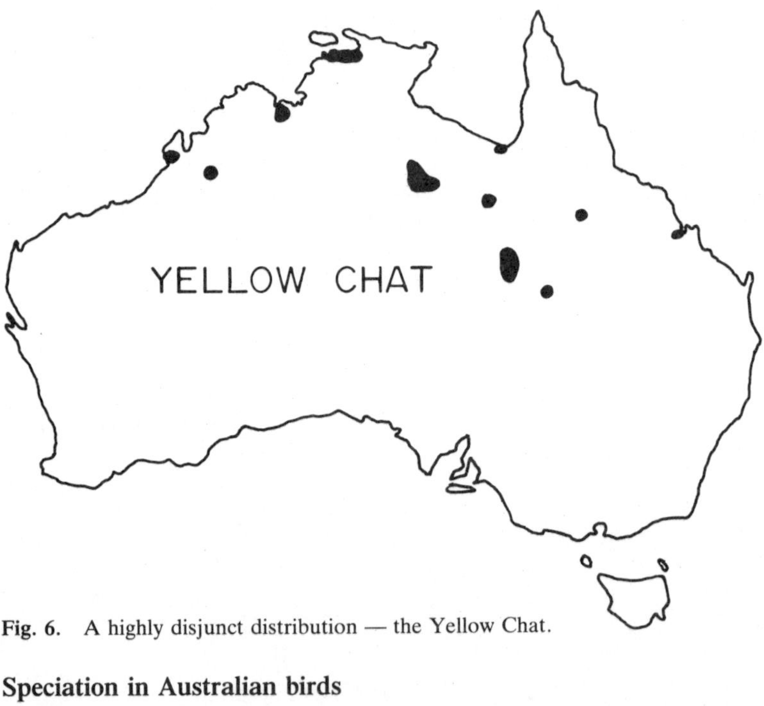

Fig. 6. A highly disjunct distribution — the Yellow Chat.

Speciation in Australian birds

Speciation should be seen principally as an ecological phenomenon. In the classical (allopatric) model, populations become isolated from each other and then diverge (Mayr 1942).

In very small populations, such as occur for instance on oceanic islands, changes may occur through chance or through genetic revolutions (Templeton 1980). However, the majority of changes are in response to selective pressures imposed by the local environment (Ehrlich *et al.* 1974). All of Tasmania's endemic species have close relatives on the mainland and in a few cases the changes that they have undergone can be explained. The Strong-billed Honeyeater (*Melithreptus validirostris*), as its name suggests, has a powerful bill, which it uses to extract insects from deep crevices in bark. Tasmania lacks bark-feeding birds like Sittellas (*Daphoenositta*), treecreepers (*Climacteris*) and Shrike-tits (*Falcunculus*). This provided an opportunity for a honeyeater to fill this role (Keast 1968a). Island birds also tend to be duller than their mainland relatives. This is probably because, with fewer species of birds on the island, there is less likelihood of confusion in mate-choice and consequent interspecific hybridization (which nearly always results in no young). The Forty-spotted Pardalote and Dusky Robin (*Melanodryas vittata*) are both duller than their likely ancestors, the Spotted Pardalote (*P. punctatus*) and Hooded Robin (*M. cucullata*) respectively.

Just as sea-barriers provide opportunities for allopatric speciation, so do arid barriers or, in the case of many Northern Hemisphere birds, glaciation during ice ages. One can see, in a range of species, the whole succession from isolated populations that are identical, through those with slight, then marked differences to two clearly distinct species. This can be best illustrated with a variety of parrots (Cain 1955). Purple-crowned Lorikeets (*Glossopsitta porphyrocephala*) appear identical morphologically in southwestern and southeastern Australia. In the Regent Parrot there are slight morphological differences between east and west. The Ring-necks (*Barnardius*) provide a more complicated pattern (Figure 7). The Mallee Ringneck (*B. barnardi*) in the east has a green head with a red forehead, southwestern birds have a black head and red forehead (Twenty-eight Parrot – *B. zonarius semitorquatus*), whereas those in between (Port Lincoln Parrot – *B. z. zonarius*) have a black head and no red forehead. These three overlap and hybridize so that it is hard to decide whether they are true species or merely subspecies. There is also a northern subspecies, which is paler and smaller (*B. b. macgillivrayi* – the Cloncurry Parrot). The Western (*Platycerus icterotis*) and Eastern Rosellas (P. *eximius*) are clearly distinct species separated by nearly 2,000 km of dry country (Figure 8). This group of rosellas is particularly interesting as it also has northeastern (Pale-headed

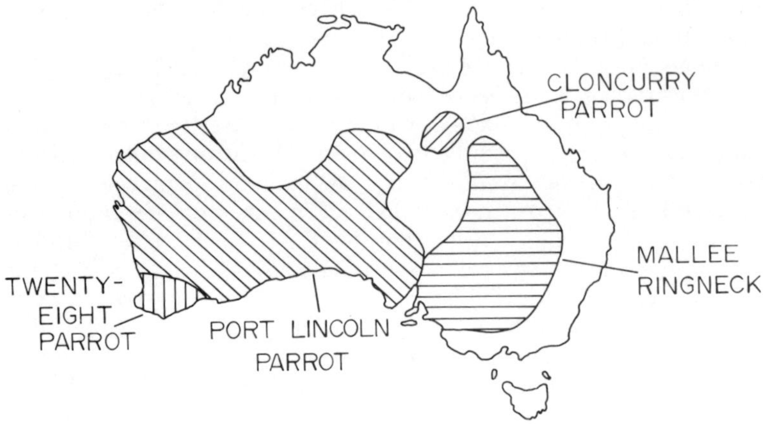

Fig. 7. Distributions of differentiated populations of Ring-neck Parrots.

Rosella – *P. adscitus*) and northwestern (Northern Rosella – *P. venustus*) representatives. Although they differ greatly in plumage, *P. eximius* and *P. adscitus* will hybridize where they overlap. Finally, the Red-capped Parrot (*Purpureicephalus spurius*) of the southwest, is so distinct from other parrots that it is placed in a genus of its own.

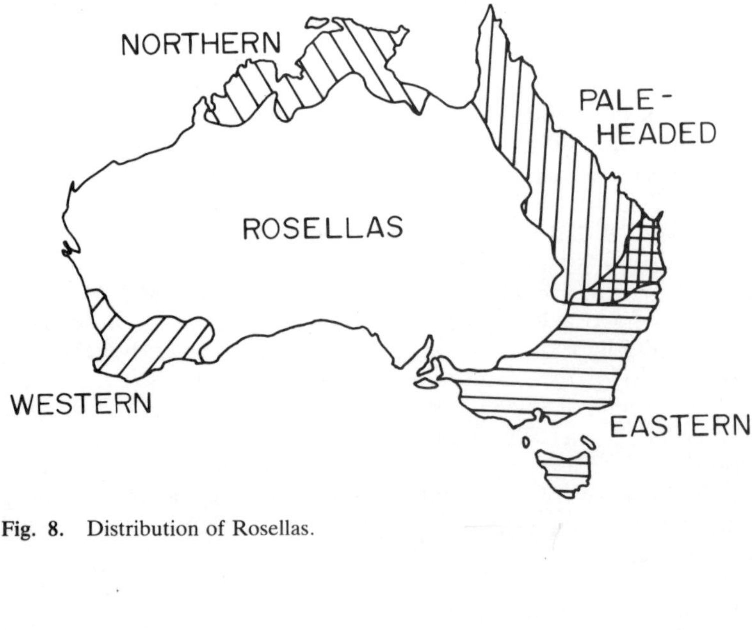

Fig. 8. Distribution of Rosellas.

Although occasional crossing of barriers may give rise to isolated populations, more often isolation results from contractions of range as the environment deteriorates. This probably happened repeatedly during the Pleistocene, as the interglacial wetter periods allowed species to spread across or around the Nullarbor Plain whereas the dry glacials eliminated intervening populations (Keast 1961). Often it is hard to provide ecological reasons for the changes that have taken place in isolated populations. I know of no explanations why Eastern Rosellas have white cheeks and Western Rosellas yellow cheeks. Occasionally, though, changes can be interpreted ecologically. Southwestern Australia has black cockatoos with white tail patches, which were regarded to be specifically distinct from yellow-tailed birds in the southeast, from which they were isolated. Saunders (1974) however found that the picture was more complicated. There are two distinct populations in the southwest. One, Baudin's Black-Cockatoo (*Calyptochynchus baudinii*), occurs in the extreme southwest and has a long beak, which it uses to extract seeds of the marri (*Eucalyptus calophylla*). The other, Carnaby's Cockatoo (*C. funereus latirostris*), occurs further north, and has a shorter, heavier beak. It eats seeds from the large woody fruits of *Dryandra* and *Hakea*. Saunders suggested that the two forms evolved alongside each other in neighbouring habitats, possibly isolated by a belt of country that provided few nest hollows. In relative isolation they specialized on different types of food. Later Saunders (1979) proposed that *C. baudinii* resulted from a population that had been cut off during an early dry period, whereas *C. funereus latirostris* originated from a much more recent colonization by the Yellow-tailed Black-Cockatoo (*C. funereus*) from eastern Australia.

Although I have concentrated on changes taking place between isolated populations, slight changes may also occur along continuous populations. These are called clines (Endler 1977). Two patterns are so frequent that they have been termed biogeographic rules. Bergmann's rule states that members of a species tend to be larger in the cooler parts of their range. For homeothermic (warm-blooded) species, this could be because heat-loss is relatively lower from a larger body. Alternatively, larger-bodied animals may be able to withstand periodic food shortages better than smaller ones. In Australia the Willie Wagtail is approximately 11% larger at latitude 38° than it is at latitude 12° (Keast 1961). Gloger's rule states that members of a species tend to be paler in the drier parts of their range. The Yellow-throated Miner

(Manorina flavigula) is far paler in central Australia than in the moister extreme southwest and southeast of its range. Indeed one population in the southeast, which may have been isolated previously, is recognized as a distinct species (Black-eared Miner *M. melanotis*) on account of its darkness (Serventy 1953). Reasons for Gloger's rule are unclear though perhaps paler birds reflect solar radiation better than darker birds do. This would allow them to feed for longer on hot days. Such an explanation would not hold for the Australian Magpie (*Gymnorhina tibicen*) where the black-backed forms occur north of the white-backed ones (Burton and Martin 1976).

Clines may result from natural selection operating differently in different parts of a species' range. Where there is a steep gradient in selection a species may show an abrupt change in a character, even without isolation. Such stepped clines are very hard to distinguish from secondary hybridization where two previously isolated forms come together and interbreed.

So far I have concentrated on species that occur as isolated populations in rather moist habitats separated by dry intervening habitats. But what of the desert species? The grasswrens (*Amytornis*) are basically a desert group whose members form isolated populations across most of inland Australia (Keast 1958a). They occupy patchily distributed habitats, mostly spinifex (*Triodia*), in sandy or rocky areas (see Plate 2). As grasswrens are very poor fliers, populations become isolated by unsuitable habitats such as salt lakes, gibber plains, mulga scrub and grassland. Presumably the spinifex and other habitats have alternately contracted and expanded during recent dry and moist phases. During expansion new areas were colonized and during contraction these populations were isolated again. The grasswrens are particularly interesting because, as Keast showed, many of the populations are obviously adapted to local conditions. Some species are reddish brown and striated, which makes them hard to see among the spinifex on sandhills that they occupy. Those inhabiting dark sandstone gorges are dark brown. Also some species have thick bills and consume many seeds, in contrast to the majority of species, which are thin-billed, and chiefly insectivorous.

Distribution, Food and Habitat

It is rarely possible to link the distribution of a bird species in Australia to one or a few major types of food. Insectivorous species and raptors are mostly very generalized in their food, varying it from place to place, but there are exceptions. Letter-winged Kites (*Elanus scriptus*) are most common in the range of the plague rat (*Rattus villosissimus*), and breed best when rats are abundant. Herbivores show a more marked tendency to match the distribution of food types. Fig Parrots (*Psittaculirostris diophthalma*) feed principally on figs and are restricted to the three extensive areas of rainforest where figs are common: Cape York, the Atherton region, and the Border Ranges. They are now extremely rare in the last area, despite an abundance of figs. Other frugivores take a wider range of food, but many are restricted to areas where fleshy fruits are present for most of the year. New Holland and White-cheeked Honeyeaters (*Phylidonyris novaehollandiae* and *P. nigra*) occur mostly in coastal banksia heathland, but like most honeyeaters they visit a wide variety of plant species for nectar (Recher and Abbott 1970; Paton 1985a).

It is often hard to separate the influence of food from the influence of habitat on the distribution of bird species. Also, ornithologists use the term habitat very loosely. Most people appreciate what is meant by mangrove or rainforest, but semi-arid woodland or desert are far looser terms. The terms mulga and saltbush plains imply habitats where these plant species are dominant. However, it is not clear whether birds occupying such habitats are attracted by the plant species or by the structure of the vegetation. To take an example; grasswrens were described previously as preferring spinifex grassland. Although they may eat seeds of the spinifex it is probably the year-round cover provided by the plant that determines their choice of and survival within the habitat. In other cases species may be forced into habitats to which they are not primarily adapted. J. Ford (1982) discussed the mangrove specialists in northern Australia and suggested that many of these originally inhabited rainforest but retreated into mangroves as aridity eliminated the rainforests. Other species occupy a wide range of habitats, and a few even change from one habitat when breeding to another in the non-breeding season.

One of the few detailed studies of habitat was that carried out by Ferrier (1985) on the Rufous Scrub-bird (*Atrichornis rufescens*). Although this species is often stated to be found mostly in cool, *Nothofagus* rainforest it in fact tends to occur either in eucalypt forest very close to *Nothofagus* or in subtropical rainforest. Ferrier measured a large number of floral and structural characteristics in Scrub-bird territories and in neighbouring sites not occupied by Scrub-birds. Scrub-birds preferred areas that had dense vegetation between 0 and 1 m above the ground. This is probably both because these birds are very secretive, and also because these habitats produce a high humidity, plentiful leaf litter and perhaps abundant food. It did not seem to matter what plants made up the understorey or indeed whether they were alive or dead. More painstaking studies of this type are needed in order to define accurately a species' habitat and dispense with loose terms such as woodland and desert. One context in which a species' habitat is particularly interesting is where two or more species have abutting ranges yet occupy different habitats. The quail-thrushes illustrate this well (J. Ford 1983). The Spotted Quail-thrush (*Cinclosoma punctatum*) occupies dry sclerophyll forest along the Great Dividing Range from central Queensland to southeastern South Australia (Figure 9). The Chestnut Quail-thrush (*C. castanotum*) occurs inland on sandy substrates in

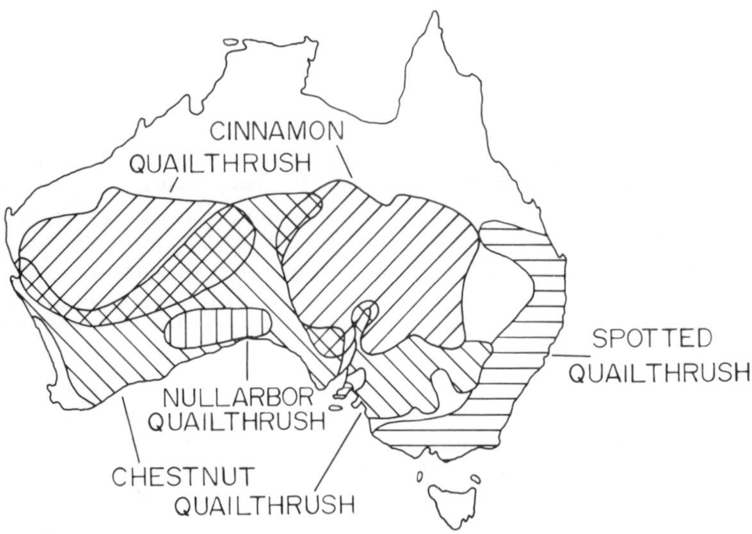

Fig. 9. Distribution of Quail-thrushes.

habitats such as mallee and native pine (*Callitris*) woodland. The Cinnamon Quail-thrush (*C. cinnamomeum*) also occurs inland and in several places overlaps with the Chestnut Quail-thrush. This species prefers stony substrates, either with mulga scrub or with saltbush. This separation is shown best on the Nullarbor Plain, which is covered in saltbush and has only a distinct subspecies of *C. cinnamomeum*, the Nullarbor Quail-thrush (*C. c. alisteri*). It is almost totally surrounded by Chestnut Quail-thrushes which occupy coastal sand-dunes and the Great Victoria Desert.

Perhaps the best example of separation of closely related species by habitat is shown by the honeyeater genus *Meliphaga (Lichenostomus)*. Keast (1961) commented on the distinct habitats occupied by some members of this genus in eastern Australia. Ford and Paton (1976a) provided a more detailed picture for South Australia, whereas Wykes (1982) quantified the habitats of several Victorian species. The clear separation is perhaps more remarkable because the species are so similar in morphology and in their foraging behaviour. Also, many of them are rather mobile and may occur outside their normal habitat in the non-breeding season. It seems hard to believe that each species is unable to occupy other habitats. Consequently it has been proposed that the distributions are kept distinct through interspecific competition (Ford and Paton 1976a). Honeyeaters are aggressive birds, not only to conspecifics. The competition hypothesis is supported by the observation that where a particular species is absent its characteristic habitat is occupied by another species. The Yellow-faced Honeyeater (*L. chrysops*) typically breeds in eucalypt forest in eastern Australia. It is absent from Kangaroo Island and south-western Australia. Purple-gaped (*L. cratitia*) and Singing Honeyeaters (*L. virescens*) occupy forest in these areas, even though both typically live in drier habitats in eastern Australia. The hypothesis that distributions are restricted by interspecific competition can only really be tested by removing one species where its habitat abuts that of another or adding species to appropriate habitat where it does not already occur. Such experiments, in addition to being ethically undesirable, are difficult with birds because they are so mobile.

Interspecific competition has been proposed as a major factor in determining the distribution of birds on islands, particularly small islands in archipelagos (e.g., Darwin's finches of the Galapagos). Diamond (1975a) plotted the distributions of bird species in the Bismarck Archipelago, off New Guinea. Several species were only found on small islands with depauperate avifaunas. These he termed super-tramps. Many of them had corresponding congeners on larger islands, but never did the two occur together on the same island. For instance, the honeyeater *Myzomela pammelaena* and the cuckoo-dove *Macropygia mackinlayi* are characteristic of small islands, but are absent from the large ones, which have other *Myzomela* and *Macropygia* species. Similar patterns are found among pigeons and hummingbirds in the West Indies (Lack 1976). Patterns of distribution such as these provide persuasive evidence for competitive exclusion. However there are alternative explanations. The eastern Polynesian islands have two species of fruit-dove (*Ducula*) showing a checkerboard distribution. Holyoak and Thibault (1978) have noted that *Ducula latrons* only occurs on islands where the Fiji Goshawk (*Accipiter rufitorques*) is absent. It appears equally likely that it is excluded by predators, rather than by competition. Again, experimentation could test these alternative hypotheses, though detailed measurement of food availability and diet could also be carried out.

Patchy distributions are not restricted to species found on islands. J. Ford (1982) noted that mangrove vegetation along the northern Australian coast consisted of small islands of habitat. Gaps occurred in the range of nearly all the mangrove specialists. In some cases absences of warblers (*Gerygone*) or whistlers (*Pachycephala*, *Colluricincla*) can be explained by competition from closely related species.

There is little evidence that Australian birds are restricted in distribution by natural predators. Those that occupy dense habitats probably do so at least partly because they provide a refuge from at least avian predators. Introduced predators like foxes and cats may have reduced the ranges of some ground-dwelling birds like Malleefowl (*Lepoia ocellata*) and Bush Thick-knee (*Burhinus magnirostris*), though the evidence is only circumstantial.

Distribution and Climate

Although biological features such as food, predators, competitors and habitat, play a major part in limiting the distribution of bird species, climatic limitations may also be important. In Australia the most likely limitations are heat and aridity. Birds can usually tolerate air temperatures up to about 45°C (Dawson and Bartholomew 1968). When it gets hotter than this birds may retreat into burrows where these are available (Attwood 1982; Ambrose 1984). They also may use evaporative cooling, with gular fluttering, though this increases their requirement for water. Adaptations to hot, dry conditions have been studied in Australia, specially in the southwest. Williams and Main (1976) found that the more coastal White-fronted Chat (*Ephthianura albifrons*) lost far more water through evaporation and had a higher metabolic rate than did the desert-dwelling Orange Chat (*E. aurifrons*).

Differences may be found even within a species. Ambrose and Bradshaw (1988a,b) found that White-browed Scrub-wrens (*Sericornis frontalis*) at Shark Bay, which is hot and dry, conserved water better and tolerated more saline water than those at a cooler, moister site farther south. Thus, small birds in Australia can tolerate hot and dry conditions in a variety of ways. There are limits, however, and occasionally very high mortality occurs (McGilp 1932).

Most species of birds avoid the extremes by migratory or nomadic movements. These will be discussed in Chapter 9. The few small, apparently sedentary species, such as the Gibberbird (*Ashbyia lovensis*) and the grasswrens, would be ideal species in which to study the response to extreme heat and water deprivation.

Summary

The distributions of birds in Australia appear to be limited by similar factors to those that limit species in the Northern Hemisphere. However, water barriers and high mountain ranges are relatively unimportant, whereas arid regions are much more important. Habitat can be used to explain the distributions of many species, but only rarely have the exact requirements of a species been elucidated. Food and competitors may limit the range of some species, but most species are generalized feeders, and there is no experimental proof of interspecific competition where ranges of two species abut. Predators would appear not to be important, though introduced predators may have reduced the ranges of a few species.

Distributions of birds on archipelagos of the southwest Pacific often show mutually exclusive patterns which have generated interesting hypotheses. Dispersal ability and competitive exclusion could account for many of the patterns. These hypotheses should now be tested by detailed field work and, if possible, experimentation.

3. Food and Foraging Behaviour

Birds eat a wide variety of plant and animal foods. The most frequently consumed items are insects, seeds, fruit and nectar. There is a specialized terminology relating to feeding behaviour and ecology. *Herbivory*[1] refers to the eating of plants and includes many categories, the most important ones for birds being: the eating of seeds (*granivory*), consumption of fruits (*frugivory*) and feeding on nectar (*nectarivory*). When the prey is killed and consumed *predation* occurs; if the host is invaded or partially consumed but usually not killed it is called *parasitism*. Many birds eat insects (*insectivores*) or vertebrates (*raptors*). Very few birds are parasites, but some Darwin's finches consume the blood of seabirds, obtained by pecking at the base of the wing or tail feathers (Bowman and Billeb 1965).

Optimal Foraging

A considerable body of theory, generally supported by observations and experiments, has developed around the apparent decisions that animals need to make about food (Krebs 1978). A primary question regards the items to be eaten. Some foods are more profitable than others, because they are larger, richer in energy or easier to handle. These should be preferred. When the best foods are scarce others need to be added to the diet. On the whole birds tend to expand their diets in seasons when food is relatively scarce. Exceptions may result from an overall reduction in the variety of foods available or from severe competition with other species.

Species differ in their preferred foods because they display different morphological adaptations, in particular in the shapes of their beaks. The larger finches in both Europe and the Galapagos (Newton 1972, Grant 1986) are more efficient at husking large and hard seeds than are smaller species.

Birds can increase their efficiency when foraging by using specific searching tactics. They may seek particular items,

[1]Terms in italics are defined in the Glossary.

developing a search image for them which increases their ability to detect them (Tinbergen 1960). They may concentrate on particular microhabitats where food is most likely to be found or in which they are morphologically adapted to forage. In uniform habitats birds may behave in ways that keep them in areas where food is abundant, for instance by making frequent turns (Smith 1974a,b). Another way to increase efficiency is to forage with other individuals of the same or different species. Successful birds can be followed and copied.

Individuals need to decide when to leave an area as food becomes depleted. Theoretically this should be when one's efficiency has declined to the average for the area (Figure 10). In practice this is difficult to determine but one can predict, for instance, that birds should move sooner when patches of food are close together than when they are widely dispersed. Studies of foraging efficiency, or optimal foraging, have had somewhat mixed success in accurately predicting what animals should do. This has been for a variety of reasons. The most profitable areas to feed may also be dangerous places. For instance, predators often visit fruiting or flowering trees. Also, birds may take food that is rich in protein or some other essential item, yet is less economical than a carbohydrate-rich food. It may be less profitable to feed on a breeding territory than elsewhere, but to leave a territory may risk losing it. Measurements of efficiency are necessarily rather crude estimates, though it is possible to be reasonably accurate with simple foods like nectar.

Also, birds are highly mobile and may find rich food sources outside the study area, and consequently desert an apparently profitable territory.

Food of Australian Birds

The most comprehensive collection of information on the food of birds in Australia is that of Lea and Gray (1935a,b,c, 1936a,b). Other significant studies include Cleland (1911-12), McKeown (1936, 1944), Green (1966) and Rose (1973). In Australia the majority of landbirds consume insects and other arthropods. Snails are eaten by the Magpie-Lark (*Grallina cyanoleuca*) and the Bassian Thrush (*Zoothera lunulata*). The main seed-eaters are the parrots, pigeons and finches, while parrots and pigeons, as well as bowerbirds, consume fruit. The con-tinent has a wide diversity of nectar-feeders, including honeyeaters, lorikeets, silvereyes and one sunbird. Lorikeets, and possibly other birds, also digest pollen

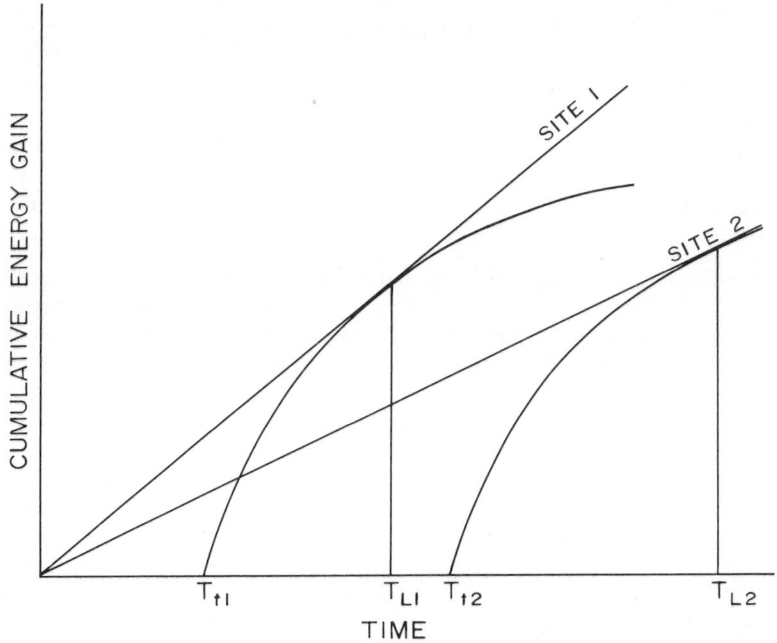

Fig. 10. One of the clearer examples of optimal foraging. Once a good area is found energy will be gained rapidly at first. As food is depleted the rate of energy gain will decline. Hence the rate of energy gain against time is a curve of declining slope. As it takes time to find a good patch of food this searching time needs to be added to calculate an overall foraging efficiency (i.e., T_{t1} and T_{t2}). The highest efficiency is shown by the straight line that touches the curve at a tangent (T_{L1} and T_{L2}). Other lines crossing the curve at other points will be less steep. Birds should leave a patch of food at points (T_{L1} and T_{L2}). It is often hard to calculate all these variables. However, qualitative predictions can be made. When patches are far apart (Site 2) the most efficient line touches the curve further to the right, (T_{L2}) i.e., a bird should spend longer in a patch.

(Churchill and Christensen 1970). One type of resource used in Australia more than on other continents is what is termed “alternative carbohydrates”. This includes sap, honeydew and other exudates from hemipterans, manna and lerp (the sugary coat of some psyllids) (Paton 1980). Vertebrates are eaten by cosmopolitan groups such as the hawks, falcons and owls, and in aquatic environments by cormorants, herons, terns and others.

The abundance of different foods varies both between habitats and seasons. Nectar is particularly abundant in sclerophyll forests and coastal heaths, whereas fruit is a major food in rainforests and succulent shrub-steppe. The finches and many of the parrots and pigeons rely on grass seeds which are more abundant in open, drier habitats. Waterbirds tend to favour swamps of shallow lakes where food is readily available. These may be ephemeral in inland Australia and may even dry up in the moister parts of the continent during droughts. Ducks and the larger wading birds therefore need to be highly mobile.

There are frequently seasonal changes in the abundance of food, and these affect the foraging behaviour of birds. Insects in southern Australia tend to peak in spring or summer and be most scarce in winter (Woinarski and Cullen 1984; Bell 1985a). However, shortages are not as severe as in Europe or North America with their colder winters. Nectar tends to be more abundant in winter and spring, though there are marked differences between habitats and years (Ford and Paton 1986). Some of the honeyeaters become more nectarivorous in winter and more insectivorous in summer (Thomas 1980; Table 3), though this is not always the case (Collins and Briffa 1982; Paton 1980). Seeds and fruits are often most abundant in summer and autumn. Finches and parrots eat more insects in winter and spring (Cannon 1981; Long 1984).

Birds also spend a greater proportion of their day feeding during the autumn and winter than at other times of the year (Table 4). This is partly because food is scarcer then, but also because days are shorter and temperatures lower, leading to increased energy demands for thermoregulation (Collins and Briffa 1983).

The foods taken may change through a 24-hour period. Honeyeaters visit flowers most efficiently in the early morning and turn to insects in the middle of the day (Paton 1982). Scarlet Robins (*Petroica multicolor*) do more aerial foraging in the middle of the day than earlier or later (Ford *et al.* 1989). Royal Spoonbills (*Platalea regia*) capture more fish and snapping shrimps at night than they do during the day (Lowe 1982). Many birds show peaks of feeding activity shortly after dawn and sometimes also in late afternoon. This may result from their demands for energy rather than the availability of food.

Different foods may also satisfy different demands. Many plant materials, especially nectar, are rich in energy but poor in

Table 3. Seasonal changes in the proportion of foraging on flowers in some honeyeaters in South Australia (Ford, unpublished).

	"Winter" April-September	"Summer" October-March
White-plumed	37%	23%
Yellow-faced	38%	20%
Brown-headed	44%	13%
White-naped	13%	2%

Table 4. Proportion of time spent foraging by various species in each season in eucalypt woodland near Armidale, New South Wales (Ford, unpublished). Number of seconds of observation in brackets.

	Summer	Autumn	Winter	Spring
Golden Whistler	—	71% (8621)	89% (8371)	48% (2772)
Rufous Whistler	43% (7443)	52% (5339)	—	33.5% (10891)
White-naped Honeyeater	72% (5140)	80% (5317)	77% (5252)	75% (1216)
Fuscous Honeyeater	51% (16948)	77% (15031)	70% (8108)	46% (12550)

protein. New Holland Honeyeaters (*Phylidonyris novaehollandiae*) in Victoria gain most of their energy from nectar, whereas insects provide their protein (Paton 1982). Indeed, it is often energetically uneconomical for them to capture insects. These honeyeaters appear to require only about 31 mg of protein per day, a very low figure compared with similar-sized birds elsewhere. Where flowering plants are well spaced, honeyeaters, as predicted by optimal foraging theory, tend to spend longer in one plant than when plants are close together (Paton and Ford 1983).

Foraging Methods and Substrates

Over the last decade many studies have quantified the foraging behaviour of Australian birds. Such behaviour is usually separated into method or technique, substrate, height and plant species used. Different methods include gleaning (picking prey off a substrate), snatching (bird in flight, prey on substrate), probing (beak at least partly inserted into substrate), pouncing (bird flies down to prey on the ground) and hawking (bird and prey in the air).

Variants of the last method include pursuing where, as in swallows and bee-eaters, the bird is in the air all the time and hovering. Shrike-tits (*Falcunculus frontatus*) and black-cockatoos (*Calyptorhynchus* spp.) tear apart bark in search of larvae and other insects. To some extent they fill the role of woodpeckers, which are absent from Australia. A significant amount of the bark-foraging is concentrated on peeling or shedding bark, which is apparently a far more important site in Australia than elsewhere (Recher 1985).

Waterbirds use another range of feeding behaviours. They dive from the surface or plunge from the air after active prey. Herons employ such techniques as leap-frog feeding, foot-stirring and paddling (Recher *et al.* 1983), whereas spoonbills sweep their beaks from side to side (Vestjens 1975).

There appear to have been very few studies of feeding rates, even though these are particularly easy to record in waterbirds like herons and waders. Recher *et al.* (1983) compared the distance moved per minute, number of steps and strikes per minute and success rates in herons.

Forest birds typically take their prey from foliage, bark, the ground or the air. Each of these categories may be subdivided, for instance foliage into that on trees and shrubs, and bark into that on twigs, branches, boughs and trunks. Other substrates like fruit, flowers and seeds, actually describe the foods being taken. Some birds display seasonal changes in the use of foraging substrate. Buff-rumped Thornbills (*Acanthiza reguloides*) and Scarlet Robins (*Petroica multicolor*) feed more on the ground in winter than in summer (Huddy 1979; Bell 1983a). There is some evidence that in winter insects are relatively more available on the ground than on bark or from the air.

The species of plant may have a major influence on birds feeding on plant material. Some eucalypt species are very attractive to honeyeaters and lorikeets, because their flowers produce large quantities of nectar. Others with tiny flowers are virtually ignored. Likewise fruit-pigeons are specialized in the plants that they visit. For insectivorous birds preferences are usually less marked. Birds are generally more abundant in forests dominated by the eucalypt subgenus *Symphyomyrtus* (most gums and boxes) than in forests dominated by the subgenus *Monocalyptus* (stringybarks, peppermints and ashes). The former grows on richer soils than the latter and Recher (1985) has suggested that *Symphyomyrtus* carries more insects and carbohydrate resources than *Monocalyptus*.

Different types of plants may also favour or disfavour different foraging techniques. The reason why few birds can live in teak plantations in New Guinea is because teak has very large leaves on which small birds have difficulty perching (Bell 1979).

Foraging Behaviour in Relation to Morphology

It is no surprise that birds' beaks are well suited to their foraging behaviour or vice versa. Honeyeaters with longer beaks tend to spend a greater proportion of their time feeding at flowers (Figure 11). Most insectivores have fine needle-like beaks (Wooller 1984), though those, like the pardalotes, that often feed

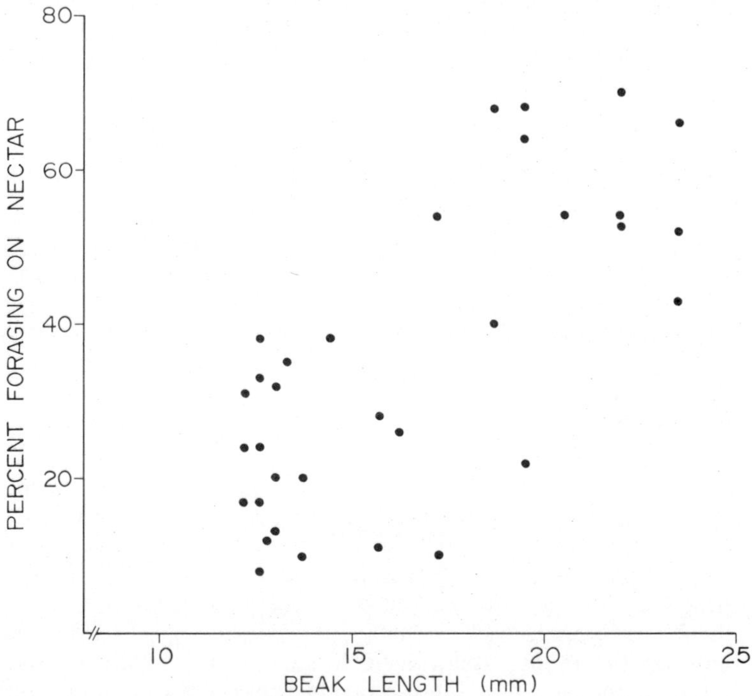

Fig. 11. Percentage of foraging observations on nectar against bill length in different honeyeaters (from Recher and Abbott 1970; Recher 1971, 1977; Ford and Paton 1976a, 1977; Thomas 1980a; Paton 1980; Ford *et al.* 1986).

on lerps have short, blunt beaks (Woinarski 1985a). Shrike-tits have massive beaks with a hook on both mandibles. Parrots too have powerful beaks, which are used for crushing seeds, and are especially effective in penetrating the hard fruits of banksias and hakeas (Saunders 1974). The waders show a diversity of beak types, which are closely related to their feeding behaviour (Dann 1987).

Other aspects of a bird's morphology are also adapted to feeding behaviour. The proportions and shapes of the wing, tail and legs all influence the efficiency of different modes of foraging (Norberg 1979). For instance, tree-climbing birds have long toes and claws but short legs, whereas aerial foragers have long wings.

Perhaps more interesting is the fact that the digestive system is related to diet. Honeyeaters have a small gizzard and short intestine compared with total insectivores (Richardson and Wooller 1986). Although they also eat many insects, those taken are usually small and soft-bodied. Frugivores also have rather simple digestive systems. The Mistletoebird (*Dicaeum hirundinaceum*) has a gizzard which is a blind sac with a sphincter. This apparently allows insects to be retained but mistletoe fruits to pass straight into the intestine (Salomonsen 1985; Richardson and Wooller 1989). Most seed-eaters have large thick-walled gizzards that grind up their hard food.

Intersexual Comparisons

Most studies on foraging behaviour of birds have been comparative ones. They have asked whether different sexes or ages of birds of the same species feed in different ways; or they have compared species, usually closely related ones, that live in the same area.

Many comparisons have involved bark-feeding birds, in particular woodpeckers, where sexual differences may be pronounced (Selander 1966; Hogstad 1976). Australian bark-feeders follow this pattern, with Sittellas (*Daphoenositta chrysoptera*) providing a particularly good example (Noske 1983, 1986). Females forage higher than males, and on smaller branches (Figure 12). These differences are related to morphological differences between the sexes. Males have longer beaks, which are more suited to probing into the deeply furrowed bark of stringybark trunks. Treecreepers, too, show intersexual differences in foraging (Noske 1986).

One of the most remarkable examples of sexual differences in foraging and morphology was shown by the extinct New Zealand Huia (*Heteralocha acutirostris*). In this bird the female had a long decurved beak whereas the male had a short stout one. Sadly, no detailed foraging study was made of the Huia, but apparently the male excavated while the female probed for insects in wood (Buller 1888).

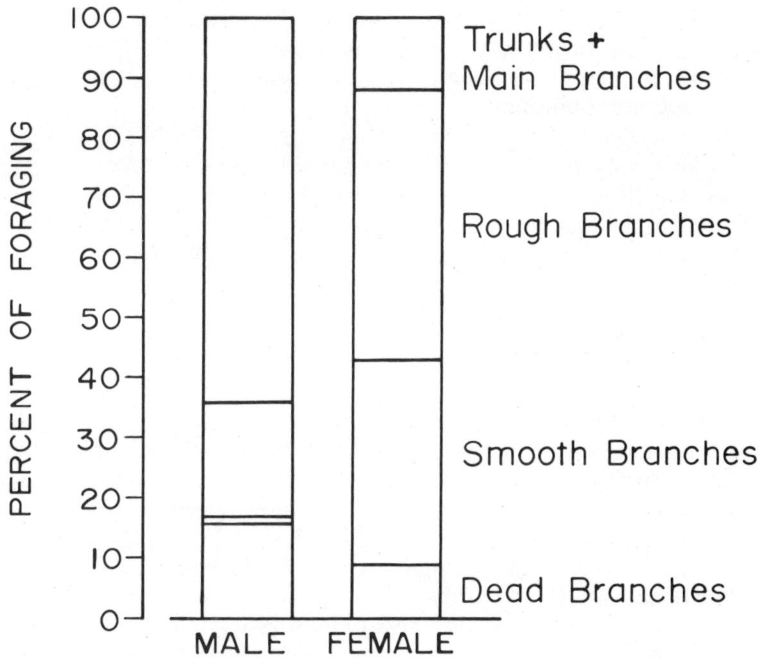

Fig. 12. Percentage of foraging by male (upper) and female (lower) Sittellas on different substrates (from Noske 1986).

Marked sexual differences in foraging are often noted to be features of depauperate environments, such as those of islands (Selander 1972). Reduced interspecific competition has allowed one species effectively to occupy two niches, the female one niche and the male another. One example though is from an island which is by no means depauperate in its avifauna. This is the Frill-necked Flycatcher (*Arses telescophthalmus*) of New Guinea (Bell 1982a). This bird occupies tropical rainforest with over 100 other bird species, of which 10 are close relatives. Male Frill-necked Flycatchers glean and snatch from leaves, trunks and branches in the understorey, whereas females glean from leaves and sally in the sub-canopy. The sexes overlap by only 58% in their foraging heights and each is more similar to three other related species. Bell suggested three possible reasons for these differences. First, there may be a shortage of true bark-feeding birds in New Guinean rainforest, which allowed the male Flycatcher to exploit a new resource. Secondly, bark-feeding may have been the original niche of both sexes, recent warming of the climate may have allowed the female to become an aerial forager. Finally, in a recent dry period the ancestors of the Frill-necked Flycatcher may have been isolated in a rainforest "island" with a depauperate avifauna.

Not only bark-feeders show sexual differences in foraging. The whistlers in Australia also differ, with males gleaning and snatching insects from higher vegetation than females and more on eucalypts than other plants such as acacias (Bridges 1980; Bell 1986a).

Interspecific Differences in Foraging

It has been known for a very long time that closely related species of birds may differ in aspects of their ecology (White 1789). One reason why this should be so is that it reduces interspecific *competition* for some limited resource and hence permits coexistence (Grinnell 1904). The Competitive Exclusion Principle states that complete competitors cannot coexist (Hardin 1960). More realistically, there should be a limit to how similar two coexisting species can be (MacArthur and Levins 1967). Numerous studies in Europe and North America have documented the ways in which similar species, typically congeners, differ in their foraging behaviours or foods (e.g., Hartley 1953; MacArthur 1958). These studies have invariably found differences, at least in the case of birds. The relevance of such studies to interspecific competition is now disputed. In particular bird populations can be controlled by factors other than food, such as predation. However, these comparative studies have provided a valuable basis for further investigations of competition, which have emphasized food depletion or experimental manipulations. (See further discussion of interspecific competition in Chapter 4.) In addition they are useful in the understanding of a species' basic requirements.

The last decade has seen a proliferation of comparative studies of foraging in Australian birds (see e.g., Keast *et al*. 1985). I shall describe two, very different, examples here: spoonbills and fantails.

Spoonbills

Spoonbills are highly specialized water-birds with only six species world-wide. Two species commonly occur together in northern and eastern Australia. Vestjens (1975) compared these at Lake Cowall in central New South Wales. The Royal Spoonbill has a shorter bill with a broader spoon than the Yellow-billed Spoonbill (*Platalea flavipes* – Table 5). It uses its bill more like a pair of tongs while the Yellow-billed Spoonbill uses its bill like a pair of forceps. Both search for prey by sweeping their bills from side to side partially submerged in shallow water. The Royal Spoonbill zig-zags its bill several times in each sweep whereas the Yellow-billed Spoonbill sweeps in a regular pendulum fashion. Once prey is located both species move the bill rapidly from side-to-side and may walk or run after the prey. Both may probe into aquatic vegetation, the Royal Spoonbill by side-to-side bill movements and the Yellow-billed Spoonbill by stabbing up and down. The Royal Spoonbill forages more rapidly and has a slightly higher success rate. Both species take a similar range of foods, with the Yellow-billed Spoonbill taking more crustaceans, such as yabbies (*Cherax destructor*) and insects like water-bugs. The Royal Spoonbill takes relatively more fish and plant material.

Table 5. Comparison of the foraging of Royal and Yellow-billed Spoonbills at Lake Cowall, New South Wales (Vestjens 1975).

	Royal Spoonbill		Yellow-billed Spoonbill	
	♂	♀	♂	♀
Bill width (mm)	180	164	231	205
Spoon width (mm)	55	51	47	46
Slow sweeping	Zig-zagging bill		Steady pendulum	
Probing	Side-to-side bill movements		Stabbing up and down	
Feeding rate	c. 56 sweeps/minute		c. 29 sweeps/minute	
Success rate	6.4 items/minute		4.7 items/minute	
Food	Relatively more fish and plant material		More crustaceans and insects	

Fantails

Fantails are small flycatchers largely restricted to the Australian region. Cameron (1985) compared three species near Kempsey in northeastern New South Wales (Table 6). The Grey Fantail

(Rhipidura fuliginosa) is the smallest but has relatively long wings and tail. It inhabits the full range of vegetation types in the study area while the Rufous Fantail (*R. rufifrons*) is found in the denser forests, where it forages in the understorey. The Willie Wagtail (*R. leucophrys*) occurs at the more open extreme,

Table 6. Comparison of the foraging behaviour of three species of Fantails (*Rhipidura*) (from Cameron 1985).

	Grey Fantail	Rufous Fantail	Willie Wagtail
Weight	8.5 g	10.4 g	20.8 g
Wing length	86 mm	87 mm	120 mm
Tail length	87 mm	83 mm	106 mm
Habitat	Wide range	Closed forest, bracken flat	Eucalypt woodland, grassland
Foraging Height	All heights	Understorey	Ground
Foraging Substrate	Sclerophyll foliage	Mesophyll and sclerophyll foliage	Ground, log, stump
Foraging Method	Hawk, hover	Hawk, hover and glean	Hawk and glean
Foraging Rate	8 perch changes/minute	14 perch changes/minute	4 perch changes/minute
Capture Rate	3 prey/minute	2.5 prey/minute	1.5 prey/minute
Food Type	Hymenoptera, flies	Flies, beetles, hymenopts	Hymenopts, beetles, flies, grasshoppers
Prey Size	3-4 mm	5 mm	7 mm
Association with other animals	Frequently in mixed flocks, attracted to humans	Occasionally in mixed flocks	Feeds from cattle, not in mixed flocks

where it forages on the ground or on sites just above the ground. All capture their prey by hawking, but display a variety of techniques to do this. Both Grey and Rufous Fantails may spin, rotating through 180° as they hop along branches. Both also flit from perch to perch, the latter more rapidly than the former, and flash their wings and fan their tails, probably to disturb insects into flight. The Grey Fantail is highly aerobatic, frequently changing direction and tumbling through the air like a falling leaf. Rufous Fantails mostly dart out in a zig-zag fashion. Willie Wagtails wag their tails from side to side and flash their wings, and pursue insects horizontally, often for some distance. In winter they spend most of their time gleaning from open ground. Grey Fantails have the highest capture rate, but take very small prey, often tiny swarming midges in winter. Willie Wagtails take the largest prey.

Rufous Fantails are summer visitors to the area, departing in March. Although some Grey Fantails and Willie Wagtails leave in autumn and others pass through, many remain through the winter. Grey Fantails often join mixed species feeding flocks of other insectivorous birds that are common in autumn and winter. Willie Wagtails do not join these flocks but often forage near cattle, even perching on them.

Morphology and foraging behaviour are related in a variety of ways in the fantails. The smallest species is the most aerobatic, whereas the largest is the most terrestrial. The Willie Wagtail has long legs suited for ground-feeding and a strong bill for coping with large and hard prey. At the opposite extreme the Grey Fantail has delicate legs and beak as it captures small, soft-bodied flying insects. The Rufous Fantail is in many ways intermediate between the other two and it is interesting that it is the one that departs in winter when food is scarce.

Insectivorous birds

Elsewhere I have attempted to draw together information on interspecific comparisons among congeners in eucalypt woodland and forest (Ford 1985a). To date there have been 12 comparisons involving 11 genera and 34 species. In most cases species have been compared in a number of characteristics and these are summarized in Table 7.

Table 7. Features in which congeneric species differ in Australian forest and woodlands (from Ford 1985a). Genera are: rosellas *Platycercus*; lorikeets *Trichoglossus*; thornbills *Acanthiza*; pardalotes *Pardalotus*; fantails *Rhipidura*; treecreepers *Climacteris*; honeyeaters *Lichenostomus* (*Meliphaga*); *Melithreptus* and *Phylidonyris*; robins *Petroica*; whistlers *Pachycephala* (Wyndham and Cannon 1985; Bell 1985a; Woinarski 1985a; Cameron 1985; Noske 1985; Wykes 1985; Ford and Paton 1976a, 1977; Keast 1968a; Recher 1977; Fleming 1980; Bridges 1980).

	Differences between congeners			
	negligible	minor	moderate	major
morphology	3	9	0	0
food*	4	6	1	0
method	2	6	4	0
substrates	2	2	6	2
habitat	2	3	5	2
plant species	1	9	2	0
height	4	5	3	0

* food not studied in robins.

Generally congeners differ only slightly in morphology and in food. This is perhaps because species need to resemble each other to be placed in the same genus by taxonomists and because similar morphology means that similar foods can be taken. In addition most birds are generalized in the foods that they eat, taking anything edible of the appropriate size. Differences in size of the bird, or particularly in its beak, may lead to somewhat different prey being taken, as in the fantails. Different species of fantails, whistlers, robins and treecreepers all differ moderately in their foraging method. Fantails differ in the amount of gleaning and hawking, whistlers in gleaning and snatching, robins in pouncing, gleaning and hawking and the treecreepers in the amount of gleaning versus probing.

Foraging substrate appears to be an important way in which close relatives differ in their ecology. This is perhaps best shown by the thornbills, where the Striated Thornbill (*Acanthiza lineata*) forages on eucalypt foliage, the Brown Thornbill (*A. pusilla*) amongst shrubs, such as acacia, and Buff-rumped Thornbills *(A. reguloides)* partition their foraging between the bark and foliage of eucalypts and the ground (Bell 1985a). Other genera differ in the amount of ground versus foliage or bark-feeding or the amount of aerial feeding compared with the use of other substrates.

As many studies were carried out in restricted areas, species that differ in habitat were not compared. In the case of these species there were invariably additional related species that were not studied but occupied different habitats. For instance the thornbill study was confined to only about 20 ha of eucalypt forest. Outside this area the Yellow-rumped Thornbill *(A. chrysorrhoa)* occupied open country, and further away the Yellow Thornbill *(A. nana)* foraged like the Striated Thornbill but in drier woodland. The fantails and the robins differed in habitat within a rather small area. The former have already been discussed but among the latter Rose Robins (*Petroica rosea*) prefer moist forests, Scarlet Robins (*P. multicolor*) dry forest whereas Flame Robins (*P. phoenicea*) are commonest in grassy woodlands.

The *Lichenostomus (Meliphaga)* honeyeaters present a classic case of congeners replacing each other by habitat. Table 8 is compiled from information from both Ford and Paton (1976a) and Wykes (1985). Despite this separation two or more species may occur together, though usually one forages in a specialized way. The White-eared Honeyeater *(L. leucotis)* is much more of a bark-forager than the other honeyeaters and frequently co-occurs with congeners. Also, several species may come together in the non-breeding season to exploit locally abundant foods like nectar. The whistlers are interesting in that they occupy the same habitat but in different seasons. Near Armidale in woodland the Rufous Whistler (*Pachycephala rufiventris*) is a summer visitor, while the Golden Whistler (*P. pectoralis*) takes over from it in April. They breed in different habitats.

Table 8. Habitat separation illustrated by the honeyeater genus *Lichenostomus* in southeastern Australia (from Ford and Paton 1976a; and Wykes 1985).

Species		Habitat
Yellow-faced	*L. chrysops*	Eucalypt forest, with dense understorey.
Fuscous	*L. fuscus*	Eucalypt forest/woodland, with sparse understorey.
White-plumed	*L. penicillatus*	Riparian woodland.
Yellow-tufted	*L. melanops*	Riparian forest, with dense understorey, or auriferous forest in rocky areas.
White-eared	*L. leucotis*	Wide range of forests, woodlands and mallee.
Purple-gaped	*L. cratitius*	Mallee-heath, dense understorey.
Yellow-plumed	*L. ornatus*	Mallee, open understorey.
Grey-fronted	*L. plumulus*	Open mallee, with *Triodia*.
Singing	*L. virescens*	Arid and coastal shrublands.

Differences in plant species or height used for foraging by congenetic bird species were generally not very great except where they related to other characteristics such as foraging substrate or habitat.

Nectar-feeding Birds

Amongst the 11 genera compared above, the chiefly insectivorous ones separate well either by foraging method, substrate or habitat, or more than one of these. In contrast the species that rely heavily on carbohydrate foods show less clear separation. Lorikeets consume nectar as their main energy source whereas pardalotes take lerp and manna. The long-beaked honeyeaters take all of these, plus honeydew. Foods such as these are scarce in the temperate forests of the Northern Hemisphere. Sapsuckers and other woodpeckers consume sap and hummingbirds take nectar, but rarely does more than one species inhabit the same place. Only in tropical areas in the Northern Hemisphere does one frequently find congeneric nectarivores co-occurring. Although they may differ in the types of flowers they visit (e.g., Snow and Snow 1972; Wolf *et al.* 1976; Feinsinger 1976) often the differences are more subtle. Feinsinger (1976) proposed that some species of hummingbirds monopolize rich sources of nectar whereas others travel between more widely dispersed flowering plants. A similar separation exists among Australian honeyeaters. Wattlebirds and New Holland Honeyeaters defend flowering shrubs and trees from smaller honeyeaters, whereas the smaller species such as spinebills (*Acanthorhynchus*) visit poorer nectar sources (Ford 1979; Paton 1979; Ford and Paton 1982). At other times nectar may be so abundant that many species can feed together with little aggression and competition. Lorikeets typically use abundant nectar sources and travel moderate distances over the year to find these. Pardalotes too are highly mobile, vary locally in numbers from year to year and could well seek areas where lerp or manna is abundant. Obviously, mobile populations are far harder to study than sedentary ones and in future the movements of lorikeets, honeyeaters and pardalotes need to be tracked in more detail.

Feeding territories

A *territory* is most simply defined as a defended area or resource (Davies 1978). Many insectivorous birds defend territories throughout the year or during the breeding season. These territories are termed breeding or all-purpose territories (or sometimes type A territories) and will be discussed in Chapter 6. Where food is localized birds may defend the resource only. Such feeding territories are common amongst nectar-feeding birds, and to some extent amongst frugivores (Pratt 1984) and waders (Myers *et al.* 1979). Territorial nectar-feeders are particularly suitable for study as nectar is quite easy to quantify. Nectar volumes and concentrations can be measured from flowers and the number of flowers in a territory can be counted or estimated. Thus one can express the value of a territory in terms of energy. Although it had long been assumed that the main function of a territory was to provide easy access to food or some other resource, this idea was put on a more scientific basis by Brown (1964), who proposed the economic defendibility hypothesis. This stated quite reasonably that there are benefits in holding a territory, but there are also costs. The benefits are in having exclusive use of a food source, whereas the costs are mostly entailed in excluding other birds by chasing, displaying and singing. One would expect animals to defend a territory only when the benefits outweigh the costs. Benefits can decline when, for instance, larger birds that cannot be removed continue to feed in the territory or when there are so many intruders that the owner is swamped. In the latter case costs also escalate. When food is scarce the benefits of exclusive use may not exceed the costs of defence. Conversely, when food is very abundant benefits from defending food may become negligible as it is just as economical to feed outside the territory.

Most work on feeding territories of nectar-feeding birds has been carried out in the tropics. One of the first attempts to quantify the costs and benefits of holding a feeding territory was Gill's and Wolf's (1975) study on the Golden-winged Sunbird *(Nectarinia reichenowi)* in East Africa. They found that territorial sunbirds spent about a quarter of their day feeding and only about 3% of their time in defence. Defence is expensive though and accounted for about 6% of their daily energy budget. The main benefit of holding a territory was that the average amount of nectar in defended flowers declined much more slowly than in undefended flowers. Flowers typically held 3–4 calories (12–17 joules) in a territory but only 1–2 calories (4–8 joules) elsewhere. Non-territorial birds would have to spend a far greater proportion of their day foraging, and may even have difficulty in balancing their daily energy budget. Put simply, the daily costs of defence are about 900 calories, whereas non-territorial birds would incur an extra cost of about 2,500 calories as they would need to double the amount of time spent feeding. Clearly territories are economical for the Golden-winged Sunbird.

It is not always profitable to defend nectar sources though. Carpenter and MacMillen (1976) studied the Iiwi (*Vestiaria coccinea*), one of the Hawaiian honeycreepers, feeding on *Metrosideros*, a tree similar to a red-flowered eucalypt. They found that the Iiwis were not territorial at very low or very high abundances of flowers (Figure 13). They were most strongly territorial at intermediate flower densities. At low abundances of flowers probably too large an area would have to be defended and this would attract too many trespassers. At high densities undefended flowers would be as rewarding as defended ones.

Australian honeyeaters may also be territorial. Paton (1979) found that New Holland Honeyeaters defended territories around a variety of native plants in Victoria. When territories were based on *Banksia marginata*, however, they were usurped by the larger Little Wattlebirds (*Anthochaera chrysoptera*) when productivity was high (Paton and Ford 1983). On the whole nectar was more abundant on territories than outside them.

A more unusual situation was studied by McFarland (1986a). He also looked at New Holland Honeyeaters, but in an upland site in northern New South Wales. The main nectar source was provided by *Banksia integrifolia* in autumn and *B. spinulosa* in winter. Productivity was highly variable and nectar was diluted by heavy rain or not produced during frosty nights. The honeyeaters had to increase their foraging time under these conditions, but sometimes were unable to satisfy their daily energy requirements. After a series of frosty nights territorial birds accumulated a considerable

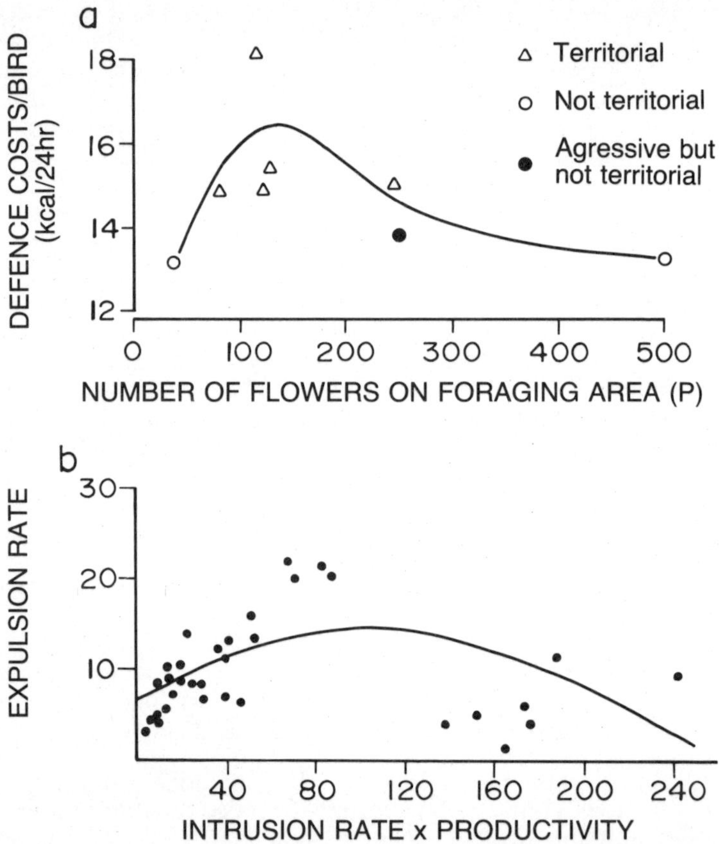

Fig. 13. Relationship between territorial behaviour and availability of nectar: a) in the Hawaiian Iiwi (Carpenter and MacMillen 1976), b) in the New Holland Honeyeater (McFarland 1986b). In both cases territorial behaviour is more intensive at intermediate levels of food.

energy deficit. What was particularly unusual was that nectar levels were lower within territories than outside them. The honeyeaters were making their territories into "nectar-sinks". Although this would not seem profitable, it may have the consequence of discouraging revisits by intruding birds. In addition the territories were probably being defended on a long-term basis, so that territorial birds could replenish their energy reserves rapidly once nectar became abundant again. Furthermore, territories are held in areas that provide prime breeding sites.

New Holland Honeyeaters are flexible in the number of intruders that they exclude. They are most diligent at intermediate productivities and intrusion rates. When nectar is very scarce or very abundant many intruders are permitted to forage without harassment. The overall pattern of territorial behaviour is compared in Hawaiian honeycreepers and New Holland Honeyeaters in Figure 13.

Other studies have found New Holland Honeyeaters to be non-territorial when nectar is abundant (Carpenter 1978) or scarce (Pyke 1983). In contrast, when nectar is very clumped in distribution New Holland Honeyeaters may show dominance hierarchies (Rooke 1979). A New Zealand honeyeater, the Bellbird (*Anthornis melanura*), also displays dominance hierarchies in the non-breeding season (Craig and Douglas 1984). Males dominate females and birds that held breeding territories near the nectar source dominate non-residents. Nectar is more abundant in the early morning than later in the day and resident males mostly feed earlier than females and non-resident males.

Honeyeaters provide abundant scope for testing hypotheses about territorial behaviour in relation to the abundance and dispersion of food. Nectar sources can be extremely concentrated and large honeyeaters such as wattlebirds and friarbirds appear to be able to switch between being non-territorial and territorial, as well as shifting to dominance hierarchies when foraging in groups. A detailed study of these large nectarivores would be valuable.

Mixed Species Feeding Flocks

Many species of birds forage in flocks for at least part of the year, sometimes only with members of their own species, though frequently with other species too. Mixed species feeding flocks, or MSFs, are found in many parts of the world (Europe: Morse 1978, North America: Morse 1970, Neotropics: Moynihan 1962, Africa: Winterbottom 1949; Greig-Smith 1978). The Australasian region is no exception. Particular species called nuclear species, form the basis of such groups and are rarely out of them. The classic examples of nuclear species in the Northern Hemisphere in autumn and winter are titmice (including chickadees — Paridae). Other species are typically attendants or followers. Most participants are insectivorous, though seed-eaters regularly join MSFs, and nectarivores and frugivores do so occasionally.

Members of MSFs may gain a variety of advantages, such as increased foraging efficiency or avoidance of predation or both. Birds may find food by watching others; even if they do not join other birds they may see where they find food. Other birds may flush insects or cause the insects to reveal themselves. By feeding in groups birds can avoid foraging over areas where food has just been depleted and also they can swamp territorial birds. The main advantage with respect to predation is that with more pairs of eyes predators are more likely to be detected. Once detected all members can be warned intentionally or otherwise and can escape or co-operate to confuse or discourage the predator. Individuals are consequently safer in a flock than by themselves. Flocks also may reduce the probability of any individual predator making contact with birds in a given area. This last aspect is obvious to any ornithologist who has searched for birds when most of them are in MSFs. Finally, by spreading the effort of being vigilant among individuals each individual can spend more time foraging.

MSFs are a common feature of eucalypt forests and woodlands (Sedgwick 1949; Bell 1980; Hermes 1981). Table 9 lists the species that frequently belong to MSFs. The chief nuclear species are thornbills, pardalotes or fairy-wrens, and these species normally occur in groups anyway. Indeed, most are co-operative breeders, which have helpers at the nest. Most of the other species are represented in the flocks either as single individuals or as pairs. Even cuckoos may join MSFs in autumn and winter, when they are not harrassed by their hosts (Bell 1986b). MSFs are most frequent between January and August, as most species do not join flocks when breeding.

Birds in rainforests in New Guinea also form MSFs (Bell 1983b), which may be present year-round. The MSFs fall into two distinct groups (Table 10). The Babbler/Pitohui assemblage is composed of large birds, most of which are a rufous colour and many mimic each others' calls. Such social mimicry has also been noted in birds in tropical America (Moynihan 1968). Clearly, such flocks have had a long history of association. The other assemblage, composed of small insectivores, is based on gerygone warblers. Most of the followers are fantails and flycatchers.

Many species alter their foraging behaviour when they join MSFs. Typically they move toward substrates where the nuclear species flock is feeding (Figure 14).

Table 9. Typical participants of Mixed Species Flocks in eucalypt woodlands and forests (from Sedgwick 1949; Bell 1980; Hermes 1981). Species are tentatively placed in two groups though some could perform both roles in different flocks.

Species	Foraging method	Foraging substrate
Nuclear species		
Buff-rumped Thornbill *Acanthiza reguloides*	Glean	Bark/ground
Striated Thornbill *A. lineata*	Glean	Leaves
Yellow-rumped Thornbill *A. chrysorrhoa*	Glean	Ground
Superb Fairy-wren *Malurus cyaneus*	Glean	Ground
Striated Pardalote *Pardalotus striatus*	Glean	Leaves
Spotted Pardalote *P. punctatus*	Glean	Leaves
Varied Sittella *Daphoenositta chrysoptera*	Glean/probe	Bark
Western Gerygone *Gerygone fusca*	Glean/snatch	Leaves
Weebill *Smicrornis brevirostris*	Glean	Leaves
Followers		
White-throated Treecreeper *Climacteris leucophaea*	Probe/glean	Bark
Golden Whistler *Pachycephala pectoralis*	Snatch/glean	Leaves
Rufous Whistler *P. rufiventris*	Snatch/glean	Leaves
Grey Fantail *Rhipidura fuliginosa*	Sally	Air
Scarlet Robin *Petroica multicolor*	Pounce	Ground
Red-capped Robin *P. goodenovii*	Pounce/hawk	Ground/air
Brown Thornbill *Acanthiza pusilla*	Glean	Leaves
Red-browed Firetail *Emblema temporalis*	Glean	Ground

Table 10. Members of Mixed Species Feeding Flocks in rainforest in New Guinea (from Bell 1983b).

Species	Colour	Size
BABBLER/PITOHUI ALLIANCE		
Nuclear species:		
Rufous Babbler *Pomatostomus isidori*	Rufous	70 g
Rusty Pitohui *Pitohui ferrugineus*	Rufous	94 g
Attendant species:		
Black Cuckoo-shrike *Coracina melaena*	♀/imm Rufous ♂ Black	55 g
Babbler Honeyeater *Timeliopsis griseigula*	Rufous	32 g
Spangled Drongo *Dicrurus hottentottus*	Black	76 g
King Bird of Paradise *Cicinnurus regius*	♀/imm Rufous ♂ Black	52 g
Raggiana Bird of Paradise *Paradisaea raggiana*	Brown & Rufous	177 g
Sooty Thicket-Fantail *Rhipidura threnothorax*	Black	18 g
SMALL INSECTIVORE ALLIANCE		
Nuclear species:		
Yellow-bellied Gerygone *Gerygone chrysogaster*		8 g
Fairy Gerygone *G. palpebrosa*		8 g
Large-billed Gerygone *G. magnirostris*		8 g
Attendant species:		
Grey-breasted Rufous Fantail *Rhipidura rufidorsa*		10 g
Chestnut-bellied Fantail *R. hyperythra*		11 g
Northern Fantail *R. rufiventris*		15 g
Black-faced Monarch *Monarcha melanopsis*		23 g
Spot-wing Monarch *M. guttula*		16 g
Black and Yellow Monarch *M. chrysomela*		14 g
Frilled Monarch *Arses telescophthalmus*		15 g
Yellow-breasted Boatbill *Machaerirhynchus flaviventer*		11 g
Olive Microeca Flycatcher *Microeca flavovirescens*		15 g
Grey Whistler *Pachycephala simplex*		21 g

The studies of MSFs in Australia and New Guinea have so far been descriptive and have not been able to discriminate between the different hypotheses that explain their evolution. Quite likely birds gain a variety of advantages from flocking and these may differ between species. For instance, most of the nuclear species forage by gleaning, often on rather exposed substrates. Perhaps they gain chiefly from the protection their followers provide from predators. For example, the Drongo is the chief sentinel of the Babbler/Pitohui assemblage. The followers forage chiefly by snatching, pouncing and hawking. They are perhaps more alert to predators but gain because the nuclear species flush or expose prey. The fantails particularly may benefit in this way.

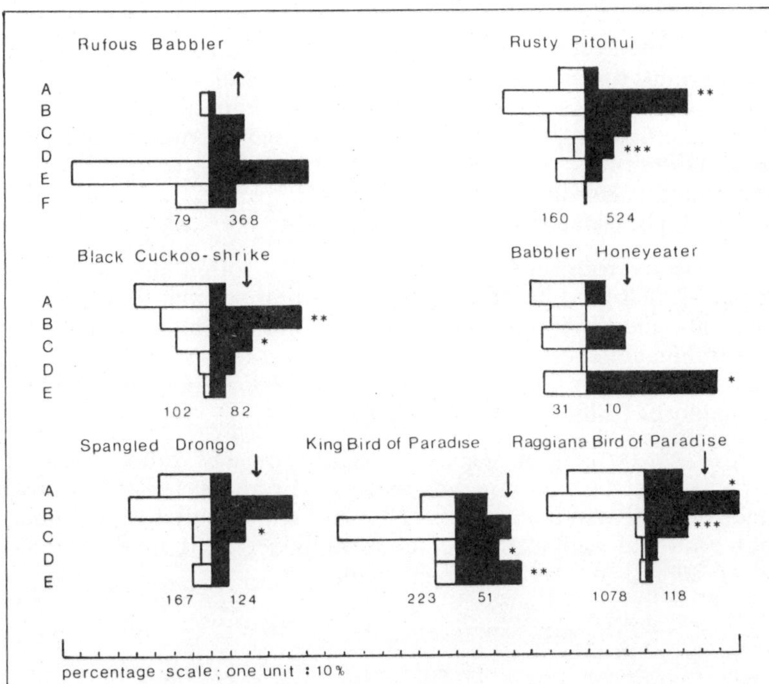

Fig. 14. Examples of changes in foraging level shown by New Guinea forest birds when they join Mixed Species Feeding Flocks (MSFs). Shaded = in MSF, unshaded = not in MSF. Arrow indicates a significant change in mean height. *, **, and *** indicate significant differences at 5%, 1%, 0.1% level, respectively in foraging when in MSF versus when not in MSF (from Bell 1983b).

In the future, studies on MSFs should concentrate on comparing the foraging efficiency of individuals in flocks with that of conspecifics foraging alone. They should also experiment with mounts or models of predators to see how quickly groups of different sizes detect them.

Summary

Birds eat a wide variety of food. Individuals tend to eat foods that are common and to which they are morphologically adapted and can exploit efficiently. Birds have frequently been used as subjects to investigate the theories of optimal foraging. Their behaviour tends to agree with predictions, though they do not necessarily attempt to maximize foraging efficiency, e.g., where there is a risk of predation or where the need for protein or other nutrients is overriding.

In some species the sexes differ in their foraging behaviours. This seems to be especially true of those that forage on bark. Numerous studies, including many Australian ones have compared the foraging behaviour of related species living in the same area. Differences are invariably found, particularly in the foraging substrates or habitats occupied. However, overlaps among species can be high, especially if food is abundant.

Insectivorous and nectarivorous birds often defend territories, which may be areas of habitat or of specific foods. Conspecifics and sometimes other species are excluded. Work on Australian honeyeaters, in agreement with studies elsewhere, indicates that territorial behaviour is most vigorous at intermediate abundances of food.

Many birds join flocks of their own and other species, especially in the non-breeding season. Mixed species flocks have been documented in Australia and New Guinea, but more detailed observational and experimental work is required to determine what advantages the participants gain.

4. Community Ecology

A *community* consists of the species or populations within a defined area (Odum 1959). It refers to all living organisms: animals, plants and micro-organisms, but often the term has been used incorrectly to refer to single taxa, as in plant "communities", or bird "communities". *Assemblage* is the correct term to describe collections of species of a particular taxon.

In this chapter I shall discuss the role that birds play in communities. Then I discuss the composition and structure of bird assemblages. I shall address the questions of what features affect the diversity and organization of bird assemblages. Finally I shall discuss change and stability in bird assemblages.

The Role of Birds in the Community

Members of communities interact with each other in a variety of ways. Some species eat members of other species. Working out which species eat which others leads to the construction of a food web. A simplified food web is shown on Figure 15. Clearly, in

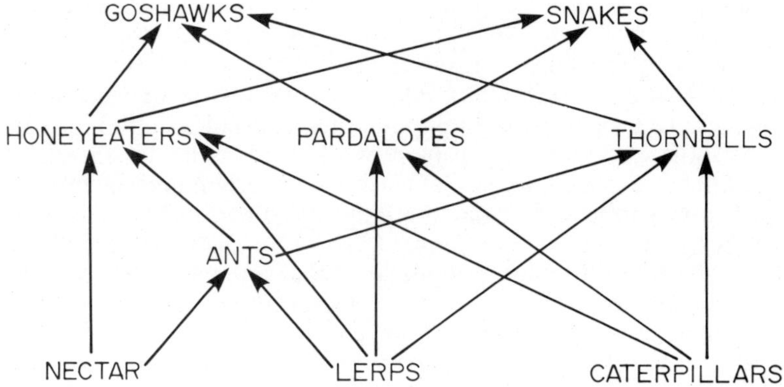

Fig. 15. An example of a food web. Arrows indicate direction of energy flow.

most communities birds can be predators of some species and prey of others.

Species may harm each other without predation, if for instance they deplete common food sources or prevent each other from obtaining critical resources like food or nest-sites. This is *competition*, a process that is believed to play an important part in structuring bird assemblages (e.g., Cody 1974). Indeed work on birds was important in the early development of ideas on competition in communities.

On other occasions species may benefit each other; this is termed *mutualism*. Familiar examples include the fungi and algae that together constitute lichens and the pollination of plants by bees.

Role of Birds as Predators of Herbivorous Insects

One role of birds in communities that has received much attention is that of their consumption of insects in forests. Forests, and most other plant communities, periodically suffer devastating outbreaks of herbivorous insects, typically caterpillars, beetles or sawfly larvae. Birds, along with predatory insects, spiders, mammals and reptiles, prey upon insects. Otvos (1979) reviewed the effects of insectivorous birds in forests and concluded that they have a significant influence on the population dynamics of many forest insects. Such an effect is hard to prove, since evidence needs to be collected on many aspects over a long period. Birds can consume 80% or more of a population of herbivorous insects, though this may not necessarily lead to control. Efforts to increase bird numbers have on occasion led to fewer outbreaks of insects (reviewed by Bruns 1960). Birds may affect herbivorous insects indirectly by spreading viruses and other diseases (Entwistle *et al.* 1977). However, outbreaks still occur and are evidence for the periodic lack of control on populations, other than ultimately by food shortage. Although far more work needs to be done it seems reasonable to conclude that, at least in northern temperate forests, birds may suppress or delay the build-up of insects from endemic levels to outbreak proportions (Otvos 1979). Some workers have even attempted to place a monetary value on insectivorous birds. For instance Evening Grosbeaks (*Hesperiphona vespertina*) were estimated to be worth over $1,000/km^2 in a western U.S. coniferous forest (Takekawa and Garton 1984).

Australian forests and woodlands also suffer defoliation by

insects. The major herbivores are scarabs (Scarabeidae) and leaf beetles (Chrysomelidae), various moth larvae, sawfly larvae (Pergidae), stick-insects (Phasmida) and sap-sucking bugs (Psyllidae, Coccidae) (Old *et al.* 1979). In forests the trees mostly recover from defoliation and continue to grow. In woodlands, however, especially those subjected to grazing, repeated insect defoliation may lead to widespread death of trees, in particular eucalypts.

Many of the prime grazing areas of Australia (for instance, the Northern Tablelands of New South Wales) are now suffering from Eucalyptus Dieback. Dieback is caused by a complex interaction of factors, including partial clearing of trees, altered microclimate, changed soil structure and composition, and introduced grasses, but the most obvious symptom is defoliation by insects (Heatwole and Lowman 1986). Trees are repeatedly stripped of their foliage, until their energy reserves are depleted and they can no longer produce new shoots. Birds have been suggested as important predators of these defoliators, e.g., scarabs (Vestjens and Carrick 1974), psyllids (Clark 1964) and stick insects (Readshaw 1968). Indeed Clark and Readshaw suggested birds could control local outbreaks of psyllids and stick-insects. Overall, birds may consume about 30 kg (wet weight) of insects per hectare per year (Ford 1986a). In years when insect populations have not broken out, this may represent 50% or more of the insects produced. Further evidence of the effect of bird predation on insects in eucalypt woodland is presented in Figure 16. The numbers of insects on branches protected from birds are higher than on those to which birds have access.

A different type of experiment was carried out by Loyn *et al.* (1983). They measured the abundance of psyllids in a Bell Miner (*Manorina melanophrys*) colony in Victoria. After the miners were removed, other species of bird moved into the area and greatly reduced psyllid numbers. Bell Miners are very aggressive and chase away small insectivorous birds. In doing so they apparently maintain higher populations of psyllids, which they harvest.

Birds have been found to carry viruses that are pathogenic to moth larvae in pasture and potato crops in New Zealand and Australia (Crawford and Kalmakoff 1977; Matthiessen and Springett 1973). Whether they act as vectors of diseases of forest insects is unknown.

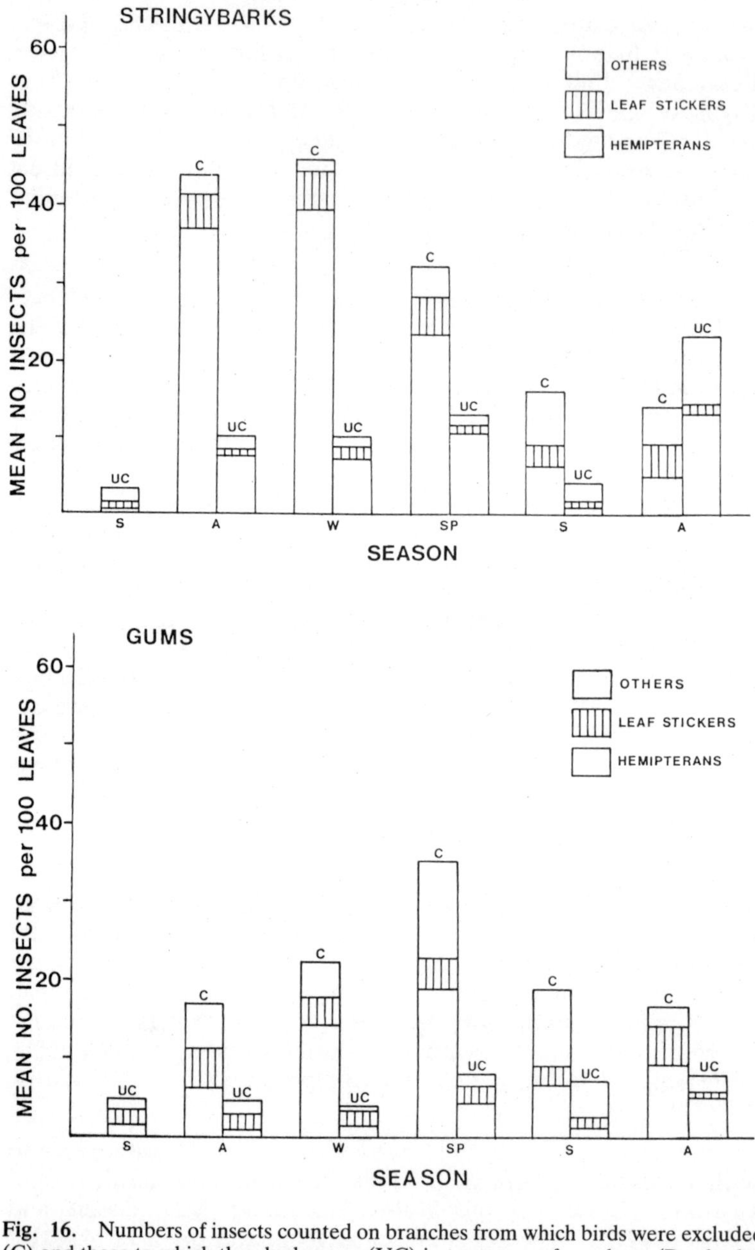

Fig. 16. Numbers of insects counted on branches from which birds were excluded (C) and those to which they had access (UC) in two types of eucalypts (Dunkerley and Bridges, unpublished).

Understanding the role that birds play or could play in regulating insect populations is undoubtedly at a very preliminary stage. There is clearly a need for more extensive research on the food of birds and how this changes as prey density alters, as well as for the monitoring of long-term changes in bird populations. The latter should be related to insect populations and other environmental changes. In addition, more should be found out about local movements and breeding success of birds. There is also scope for larger scale exclosure experiments of the type shown in Figure 16, as well as the use of nest-boxes or other habitat manipulations to attempt to increase numbers of birds.

Nectar-feeding and Pollination

Birds are far more important as pollinators in temperate communities in Australia than they are in the Northern Hemisphere. They are insignificant in Europe (Ford 1985b) and most of northern Asia, and only a few species of hummingbird venture into North America, chiefly the southwest, to pollinate plants there (Skutch 1974). It has been known for a long time that honeyeaters and other Australian birds visit flowers that are well suited for pollination by birds (Sargent 1928). However, it took detailed and careful work by David Paton to show that birds not only pick up pollen from many native Australian plants but also deposit it on the stigmas of other individuals of the correct species (Paton and Ford 1977). Plants that are pollinated by birds are surprisingly diverse (Table 11). The Epacridaceae include species that are red and tubular (see Plate 31), like many hummingbird-pollinated flowers, whereas in the Myrtaceae the flowers are usually open and cup-shaped (see Plate 32). Other pollinators including insects and mammals may visit the latter. In the Proteaceae there are gullet-shaped flowers, as in *Grevillea*, and brush-like inflorescences as in *Banksia* (see Plate 33).

Whereas David Paton showed that birds can pollinate plants Stephen Hopper and Allan Burbidge (1986) established that birds are necessary for plants to set seed. Red and green Kangaroo-paws (*Anigozanthos manglesii*) enclosed in cages set only 0.1 seeds per flower versus 22.5 seeds in flowers that had been visited by Red Wattlebirds (*Anthochaera carunculata*). It is less clear whether banksias require birds to visit flowers in order to set seed. Caged inflorescences have set as many seeds as those to which birds had access in some experiments, whereas in other cases caged inflorescences have set negligible numbers of seeds (Whelan and Burbidge

Table 11. Important genera of bird-pollinated plants in Australia (from Paton 1986a; Hopper and Burbidge 1986).

Family, Genus	Floral shape	Inflorescence	Colour*
MYRTACEAE			
Eucalyptus	Cup	Cluster	white (red)
Callistemon	Cup	Brush	red (white)
Calothamnus	Cup	Brush	red
Melaleuca	Cup	Brush	white
PROTEACEAE			
Banksia	Gullet	Brush	yellow to red
Grevillea	Gullet	Cluster	red
Adenanthos	Gullet	–	red (yellow)
Dryandra		Brush	yellow
Hakea	Gullet	Brush	red (white)
LORANTHACEAE			
Amyema	Tube	Cluster	red (green)
Lysiana	Tube	–	red (yellow)
RUTACEAE			
Correa	Tube	–	red (green)
EPACRIDACEAE			
Astroloma	Tube	–	red
Epacris	Tube	Spike	red (white)
HAEMODORACEAE			
Anigozanthos	Tube	Spike	red (yellow)
MYOPORACEAE			
Eremophila	Gullet	–	red

*Less common colours in brackets.

1980; Paton and Turner 1985). Banksias appear able to develop selectively those flowers that have received visits by birds (Vaughton 1989). Other pollinators such as insects and mammals were able to reach inflorescences in some of these experiments. In proteaceous plants, such as the Waratah (*Telopea speciosissima*), very few seeds are set by most inflorescences (Pyke 1982). This is probably not because pollinators are in short supply, but because nutrients are scarce. This could be a common problem in the infertile heaths and forests of Australia.

There have been several attempts to answer the question why Australia has so many bird-pollinated plants and nectar-feeding birds (Recher 1981; Turner 1982; Ford 1985c; Paton 1986b). The answers are highly speculative and where historical aspects are concerned they are likely to remain so.

Typically, bird-pollination in plants evolved from insect-pollination and in many cases the separation is not complete (Turner 1982 also discusses its relationship with mammal-pollination). Birds offer several advantages over insects. They are large, highly mobile and active in cold weather. This should lead to both more reliable pollination and a greater diversity of pollen arriving at individual plants. Hopper and Moran (1981) presented evidence that predominantly bird-pollinated *Eucalyptus stoatei* experiences higher levels of outcrossing than other more generalized eucalypts. This could be because birds more frequently move between trees while they are feeding than other pollinators do. Outcrossed seeds are probably competitively superior to seeds that result from self-pollination. Hopper and Moran also argued that outcrossing is particularly important for species that exist in small isolated populations, such as *E. stoatei* and *E. caesia*, so that they can maintain genetic variability. These are among the most specialized of their genus for bird-pollination.

Bird pollination is strongly associated with sclerophyllous forests and heaths, reaching its zenith in the southwest of Western Australia (Keighery 1982). This vegetation apparently originated on small patches of infertile soil within the wet forests that covered Australia in the early Tertiary (Lange 1982). Honeyeaters were diversifying from their more insectivorous ancestors at about that time (Sibley and Ahlquist 1985). Birds would have been ideal pollinators of early sclerophyllous plants, especially in the absence of advanced social bees (Apidae).

Initially plants needed to make few changes to attract birds as visitors to their flowers. The main one was an increase in the amount of nectar. Even without this, nectar could accumulate in insect-pollinated flowers if insects were scarce or inactive, for instance in cold weather (Paton 1986b). Insectivorous birds such as thornbills even now may turn to nectar when insects are scarce. Further specializations by the plants involved changes in the shape of the flowers that increased the probability of birds making contact with the anthers and stigma and reduced the visits of insects (Paton 1986b). The birds themselves require few specializations though the more nectarivorous honeyeaters have longer beaks and simpler digestive systems than do more insectivorous species (see Chapter 3).

As with the role of insectivorous birds in communities so also does the role of pollinating birds require far more study (see Ford 1986b for some suggestions). The emphasis now must be on the experimental approach pioneered by Stephen Hopper and David Paton.

Bird Assemblages

Communities and assemblages possess a series of unique properties. These include diversity, structure, succession and stability. The simplest measure of *diversity* is the number of species. This may be a poor measure for two reasons. First, the number of species depends on the area searched and time spent searching. For mobile organisms, like birds, new species will be added continually, though at a decreasing rate. Second, communities in which one or a few species dominate should be considered as less diverse than those where many species are more equal in abundance. Most measures of diversity therefore include an estimate of relative abundance. A wide variety of measures have been used and there is dispute over which is most appropriate. One of the simplest and most used is the Shannon-Weiner index: $H = -\Sigma p_i \log_e p_i$ where H = diversity and p_i = proportion of all individuals that belong to species i.

Structure deals with the composition, the identity of the species present and the niches of these species. The range of resources that a community offers determines to a large extent the species that belong to it. Another important factor is the extent to which interactions between ecologically similar species can permit or prevent their coexistence. *Succession* represents a predictable pattern of change, in a community, usually over decades or centuries. *Stability* is a complex and confused concept in ecology. To some ecologists it means lack of change, but to others it means the ability to resist change or recover from severe perturbation. I shall attempt to disentangle these various meanings.

Diversity

Many ornithologists are compulsive compilers of lists, so that information is available on numbers of species from an immense variety of sites around the globe. On a gross scale there are more species of birds as one travels from the poles to the equator in both hemispheres. A few groups go against this trend, e.g., waders in the breeding season and penguins. Tropical parts of Australia are typically richer than cool temperate regions, though productivity and isolation also influence species richness.

On a more local scale there have been attempts to relate the number of bird species, or an index of diversity, to features of the habitat. More productive terrestrial sites typically have more bird species; forests are richer than deserts (Ricklefs 1980a). MacArthur and MacArthur (1961) demonstrated that in North America diversity of bird species is more closely correlated with the structural diversity of a site (i.e., the range of vegetation layers) than to the diversity of plant species. A similar pattern has been found in Europe (e.g., Moss 1978, though see Erdelen 1984).

Recher (1969) applied the MacArthurs' method to a small number of Australian habitats and found the same pattern. The results of a much larger number of sites are given in Table 12.

Table 12. Relationship between species diversity of birds and vegetation in eucalypt forests (from Recher 1985).

	Number of Bird Species	Diversity
Foliage Height Diversity		
Mature forest (45 plots)	ns	ns
Mature and regenerating forest (69 plots)	*	*
Canopy Height		
Mature forest	ns	*
All plots	*	*
Total foliage cover		
Mature forest	ns	ns
All plots	*	*
Number species in canopy		
Mature forest	ns	*
Number of tall shrub species		
Mature forest	*	*

ns = not significant, * = Significant 5% level.

Although structural diversity is important the far simpler measure of canopy height was an equally good predictor. Within eucalypt forests the diversity of plant species in the canopy and shrub layer both contributed to the diversity of bird species. Gilmore (1985) found the opposite result, that the number of plant species was negatively correlated with the number of bird species. Recher (1985) provided an explanation for these contrary results. The relationship between bird species and plant species probably involves two main components. First, habitats with a diverse shrub layer usually include either many good nectar-producing species that attract a range of honeyeaters or shrubs of rainforest origin that attract frugivores. Second, vegetation on poor soil is often diverse in plant species but low in productivity; insects and consequently insectivorous birds are relatively scarce, and rather few in species. On the one hand, one has the floristically diverse heathlands of the Sydney sandstone and the southwest which are dominated by honeyeaters but have few insectivores. On the other hand, eucalypt woodlands are quite simple structurally and have few plant species but many insectivorous bird species. Recher's arguments are speculative but could be tested by long term monitoring of bird and insect populations.

Islands tend to have fewer species of birds and other organisms than neighbouring continents do. Britain lacks several species that breed in continental Europe at the same latitude but has several species that are missing from Ireland (Lack 1976). Generally the larger the island the more species of birds that live there (Figure 17). There are several reasons for this. A larger area

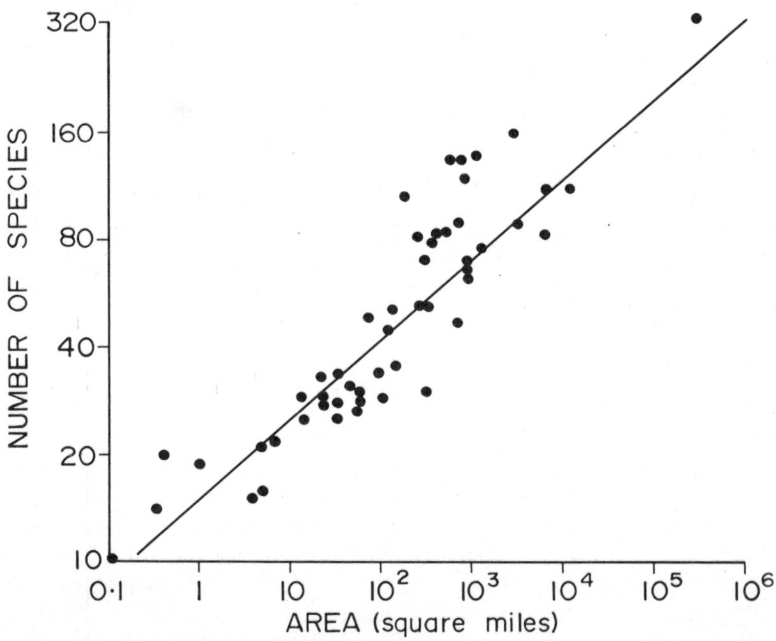

Fig. 17. The relationship between number of bird species and area for islands off New Guinea (from Diamond 1975a).

will contain more individuals and just by sampling alone these will tend to belong to more species. Some species may be absent because their habitats or some other requirements are missing.

Also, as discussed in Chapter 2, some pairs of species appear to exclude each other competitively from islands.

More distant islands also have fewer species relative to their area than islands close to continents. This may be because few species have successfully dispersed to these islands (MacArthur and Wilson 1967). Lack (1976), however, believed that the main reason was that islands are ecologically impoverished, so that a few generalists are competitively superior to many specialists.

Abbott (1973, 1974) has reviewed the avifaunas of Kangaroo Island, Tasmania and other Bass Strait Islands. They are all depauperate compared with the southern mainland of Australia. Whereas some species are apparently absent due to lack of habitat, there are no obvious reasons why others do not occur (Table 13). Detailed studies of the ecology of these species are required.

Table 13. Examples of species that do not breed in Tasmania or Kangaroo Island (i) because suitable habitat is apparently absent and (ii) despite the presence of apparently suitable habitat (from Abbott 1973, 1974; Ford and Paton 1976b).

Tasmania	Kangaroo Island
i) Suitable habitat absent or scarce	
Red-capped Robin	Eastern Rosella
Brown Songlark	Laughing Kookaburra
Yellow Thornbill	Sacred Kingfisher
	Hooded Robin
	Rufous Whistler
	White-plumed Honeyeater
	Diamond Firetail
ii) Suitable habitat present	
Rainbow Lorikeet	White-browed Babbler
Powerful Owl	Varied Sittella
Eastern Yellow Robin	White-throated Treecreeper
Jacky Winter	Yellow-faced Honeyeater
Crested Shrike-tit	White-naped Honeyeater
Striated Thornbill	
Varied Sittella	
White-throated Treecreeper	
Red-browed Firetail	

Bird Assemblages of Eucalypt Woodlands and Forests

More is known about bird assemblages in eucalypt forests and woodlands than those of any other major habitat in Australia

(Keast *et al.* 1985). Sites typically hold about 40 reasonably common species and there is little difference in diversity between northern and southern Australia (Keast 1985). There may be very marked differences in the composition of species between sites that are close together (e.g., within Victoria, Gilmore 1985; Loyn 1985). Conversely some species occur in two or more sites that are very distant. Keast (1985) sampled assemblages near Darwin and Sydney and in northeastern and southwestern Australia. Rufous Whistlers (*Pachycephala rufiventris*) and Grey Shrike-thrushes (*Colluricincla harmonica*) occurred in all four sites.

Recher and Holmes (1985) and Ford *et al.* (1986) independently recorded the foraging behaviour of bird species in eucalypt forest in southern and northern New South Wales. Their conclusions on the structure of these assemblages were very similar and are summarized in Figure 18. The species can first be separated by their main food: seeds, fruit, nectar or insects (Recher and Holmes included all honeyeaters and the Silvereye (*Zosterops lateralis*) in the nectar-feeders, whereas Ford *et al.* placed the short-beaked honeyeaters and Silvereye in the insectivores). Just over three quarters of the species are primarily insectivorous and these can be separated first by foraging substrate: ground, bark, leaves and air. The term *guild* is used for collections of species feeding on similar foods and substrates. Each guild can be subdivided by the foraging method that they use: glean, snatch or pounce. Species that were still together at this level were often congeners and the ways in which they differ in foraging were discussed in Chapter 3. Eucalypt forests and woodlands have bird assemblages with similar structures in northern and southern Australia (Keast 1985).

Probably the habitats in the Northern Hemisphere that most resemble eucalypt forest and woodland are the evergreen oak woodlands of southwestern North America and mediterranean Europe. Bird assemblages have been studied in both these habitats (Landres and MacMahon 1983; Zamora and Camacho 1984) though in less detail than in eucalypt forests. Assemblages on all three continents appear to be separated in similar ways, first by food, second by substrate and third by method. However, the proportion of species that belongs to each of these foraging guilds is markedly different in the three places (Table 14). Australia has more nectarivores, but fewer omnivores that eat insects and fruit, at least when compared with assemblages in Europe.

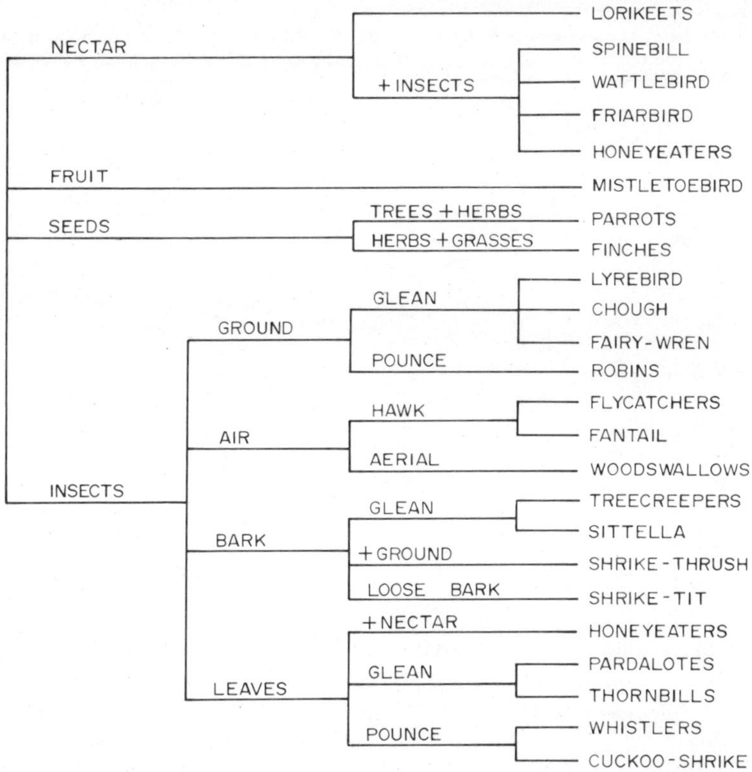

Fig. 18. Partitioning of the bird assemblage of eucalypt forest in New South Wales by food, substrate and foraging method (from Recher and Holmes 1985; Ford *et al.* 1986).

What determines the structure of bird assemblages in eucalypt forests? At the most basic level species will not be present unless there are appropriate resources for them. There are no prolific nectar-producing plants in Europe and hence no nectarivorous birds. Eucalypt forests generally have few fleshy fruits and therefore few frugivorous birds. In wet gullies there may be fruit-bearing plants of rainforest origin which attract frugivorous birds (Milledge and Recher 1985). The abundance of individuals belonging to a foraging guild is related to the amount and productivity of the appropriate substrate. Gilmore (1985) found positive correlations between the density of foliage insectivores and foliage cover; between bark-foragers and the area of bark; and between ground-foragers and annual litter fall. As the number of individuals tends to be correlated with the number of species, then a

Table 14. A comparison of the structure of bird assemblages in eucalypt forest in Australia and evergreen oak woodland in Spain and Western North America.

<table>
<tr><th></th><th colspan="2">Australia</th><th colspan="2">Spain</th><th colspan="2">North America</th></tr>
<tr><th></th><th>Ford et al. 1986</th><th>Recher et al. 1985</th><th>Zamora & Camacho 1984</th><th>Herrera 1980</th><th colspan="2">Landres & MacMahon 1983</th></tr>
<tr><th></th><th></th><th></th><th></th><th></th><th>California</th><th>Mexico</th></tr>
<tr><td>Granivore</td><td>2</td><td>2</td><td>4</td><td>4</td><td>—</td><td>—</td></tr>
<tr><td>Frugivore/ Insectivore</td><td>2</td><td>0</td><td>13</td><td>9</td><td>—</td><td>—</td></tr>
<tr><td>Nectarivore/ Insectivore</td><td>8</td><td>5</td><td>0</td><td>0</td><td>—</td><td>—</td></tr>
<tr><td>Insectivore-Aerial</td><td>6</td><td>4</td><td rowspan="4">12</td><td>6</td><td>2</td><td>3</td></tr>
<tr><td>Insectivore-Ground</td><td>9</td><td>13</td><td>4</td><td>1</td><td>2</td></tr>
<tr><td>Insectivore-Leaves</td><td>9</td><td>11</td><td>9</td><td>5</td><td>3</td></tr>
<tr><td>Insectivore-Bark</td><td>4</td><td>5</td><td>2</td><td>2</td><td>7</td></tr>
</table>

— indicates guilds not studied.

greater amount of a substrate provides opportunities for more species. The presence of a shrub layer is important for nesting sites for species such as Superb Fairy-wrens (*Malurus cyaneus*) (Nias 1984). Tree holes and mistletoes are further important resources.

Assemblage structure is therefore determined in the first place by which foraging guilds can find substrates, but what determines which members of a guild can coexist? Until recently interspecific competition has been accepted as the major governing force determining the structure of guilds (Cody 1974; Diamond 1975a). This is now being challenged (Wiens 1977, 1983) and more convincing proof of competition is being sought. I shall delay my discussion of the role of competition in Australian bird assemblages until I have summarized the guilds of nectar-feeding birds.

Bird Assemblages of Rainforests

In contrast to eucalypt forests there is a marked increase in the diversity of bird assemblages of rainforests as one goes from temperate to tropical regions (Table 15). Thomas (1980b) found that Tasmanian rainforests had very few species, most of which

Table 15. The composition of bird assemblages in four Australasian rainforests. Rare species have been excluded. Mixed feeders take invertebrates and nectar and/or fruit. Note: Thomas recorded all species; other studies were more selective.

Location	Tasmania	North Queensland Upland	North Queensland Lowland	New Guinea
Latitude	42°	19°	17°	9°
Reference	Thomas (1980b)	Frith (1984)	Crome (1978)	Bell (1984)
Herbivores				
Frugivores	0	2	6	27
Nectarivores	1	1	0	6
Granivores	1	0	0	0
Mixed Feeders	2	5	6	15
Carnivores				
Invertebrates – ground	3	6	2	11
Invertebrates – air	1	1	0	7
Invertebrates – bark	5	1	1	4
Invertebrates – foliage	7	6	12	11
Invertebrates generalized	0	0	0	2
Scavengers	1	0	0	0
Total number of species	21	22	27	83

were insectivores. In north Queensland there are more species, and frugivores are far better represented, although nectarivores are still not a major component (Crome 1978; Frith 1984). Within a guild the height at which foraging takes place appeared to separate the species best, followed by the foraging site and then the foraging method.

The most diverse assemblages of Australasia are found among the tropical rainforests of New Guinea. Bell (1984) compared the foraging ecology of 83 species in one small plot. Many more species were recorded but insufficient data were collected from them. Almost half of these species ate at least some fruit. These could be separated by their size, the level at which they foraged and the extent to which they consumed food other than fruit (Figure 19). Amongst the insectivores, the foraging substrate, size of the bird, foraging method, degree to which they took nectar and the level of foraging separated the species (Figure 20).

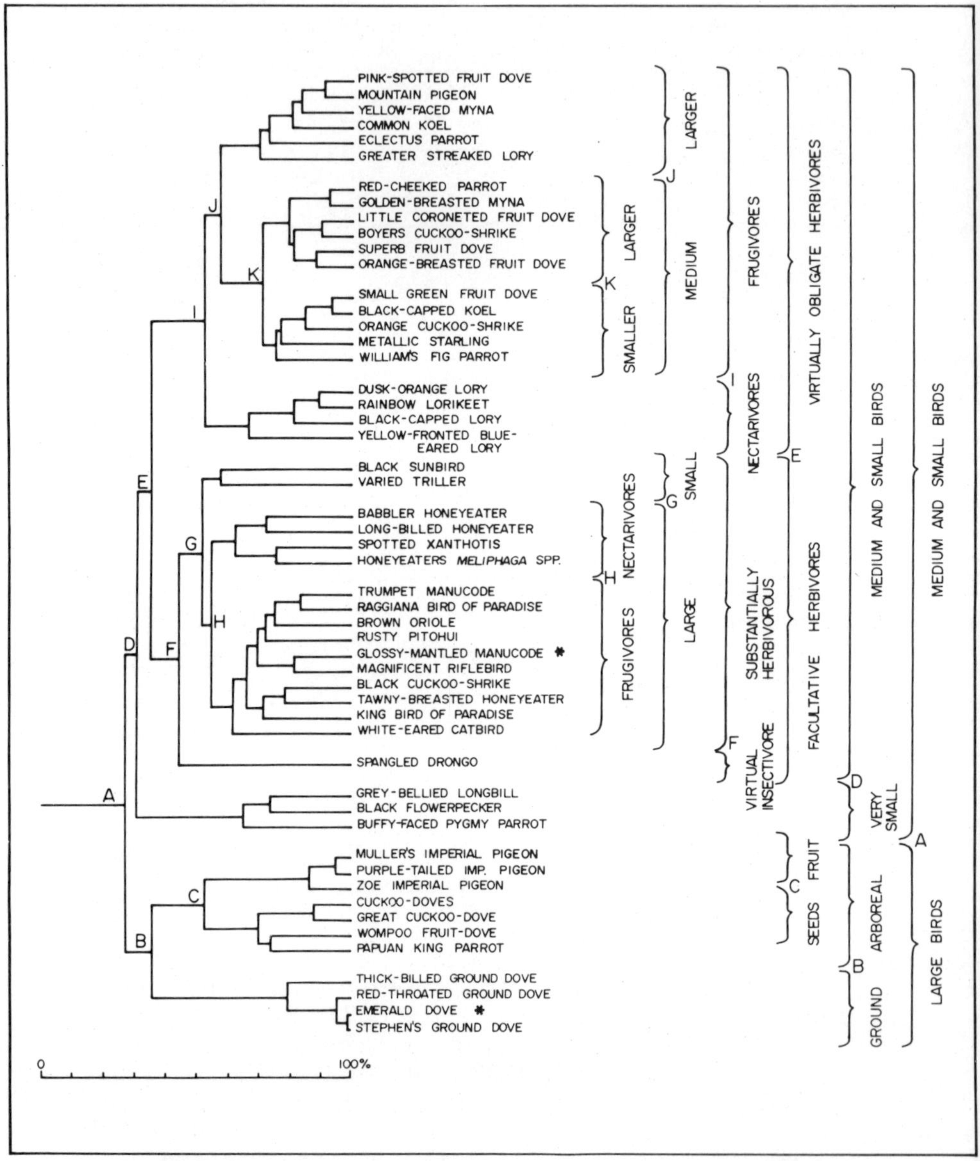

Fig. 19. Dendrogram displaying separation of herbivorous birds in rainforests in New Guinea (from Bell 1984).

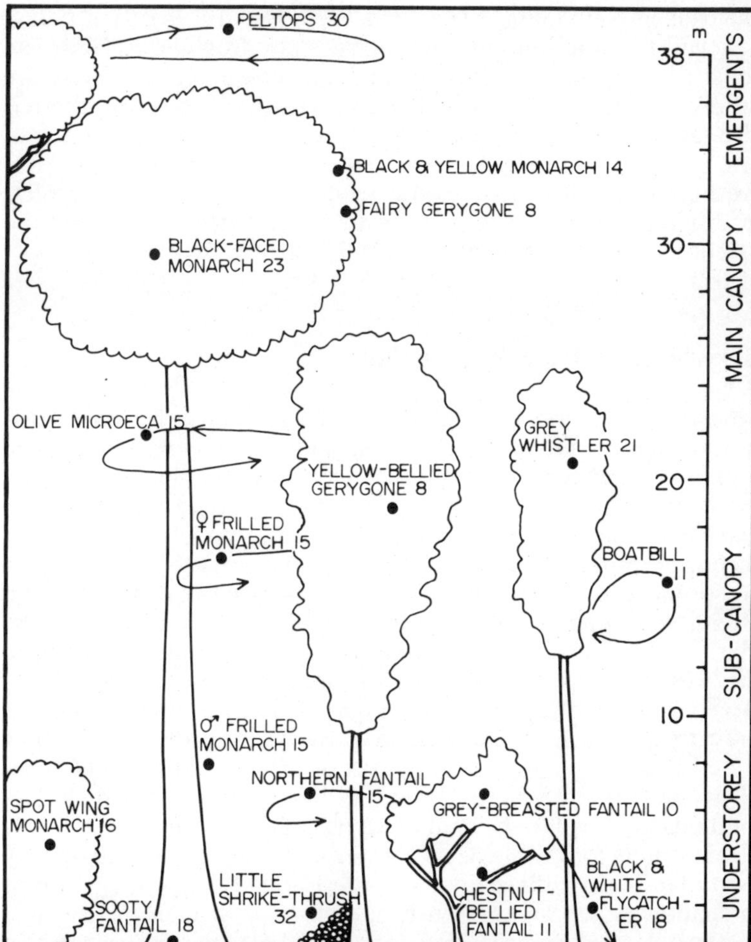

Fig. 20. Diagrammatic representation of the ways in which small insectivorous birds are ecologically segregated in rainforest in New Guinea (from Bell 1984). Numbers after each species are weights in grams.

Comparing bird assemblages in New Guinean rainforests with those elsewhere in the tropics, Bell (1984) noted that there was a scarcity of trunk-foragers but a greater diversity of frugivores. Possibly other groups of animals such as insects, lizards or mammals have taken over the trunk-foraging role of birds such as woodpeckers or treecreepers (Pearson 1977). One reason for the wealth of frugivorous birds may be the absence of mammals that eat fruit in New Guinea (Pearson 1977; Bell 1982b). Bell (1984) believed that the question of why so many frugivorous birds can coexist in one area has not been adequately answered. There are differences in sizes of fruit taken by different species, but there is also extensive overlap among species. Fruit is often abundant, removing the necessity for differences in diet and also frugivores are highly mobile. Future work needs to be directed to the diets of different species in relation to the abundance of food and the extent to which birds, and other animals, can deplete the supply of fruit.

Assemblages of Large Wading Birds

Most of the previous discussion of bird assemblages has concerned species that are typically Australian. Patterns may resemble those found on other continents but they have developed independently as the participants are different. Birds found in freshwater or marine habitats are far more cosmopolitan and hence it is less surprising to find similar structures and patterns of partitioning in Australia to those elsewhere. One assemblage that has received attention is that of large wading birds like herons and egrets. Harry Recher studied herons in North America and has compared the Australian species near Sydney and in the Northern Territory (Recher *et al.* 1983). Herons forage by standing and waiting for prey to approach or by seeking it by walking through shallow water or on the bank at varying speeds. The Great Egret (*Egretta alba*) and Intermediate Egret (*E. intermedia*) are slow and methodical, spending three-quarters of their time stationary. Great Egrets foraged in deeper water and took larger prey (mostly fish) than Intermediate Egrets. Little Egrets (*E. garzetta*) foraged in shallow water for small fish but were much more active than the two larger species. Pied Herons (*Ardea pictata*) were also active foragers but mostly stayed out of the water and apparently took insects from the vegetation. Striated Herons (*Butorides striatus*) near Sydney occupied mud flats and mangroves and stood and waited for fish in moderately deep water. The Pacific Heron (*A. pacifica*) fed in a similar way to the two large egrets but took a wider variety of prey, including insects, tadpoles, frogs and fish. White-faced Herons (*A. novaehollandiae*) walked rather quickly in shallow water in Sydney. This species is very widespread in Australia, occurs in a broad range of habitats including dry paddocks and uses a large variety of foraging methods (Lowe 1983). It tends to take rather small prey.

Most of the other heron species have been studied elsewhere. Eastern Reef Egrets (*E. sacra*) occupy estuaries, rocky coasts or islands where they stand crouching, waiting for prey (Recher and Recher 1972). Cattle Egrets (*Ardeola ibis*) typically forage in grasslands, often among large herbivores where their efficiency is enhanced (Heatwole 1965). Great-billed Herons (*Ardea sumatrana*) stand and wait, probably for large prey. The Rufous Night-heron (*Nycticorax caledonicus*) and the bitterns mostly forage at night, the latter amongst dense cover. Herons will sometimes steal prey from other birds, or rob nests; they also modify their behaviour to avoid glare from the sun (Davis 1986) and overseas have been observed attracting fish with bait (Boswall 1978).

Recher *et al.* (1983) concluded that in Australia herons differ in their foraging speed, their size and that of their prey, depth of water, habitat and time of day when they forage. They display the same range of behaviour as North American herons and partition their environment in a similar way.

Nectarivorous Birds

Ten or more species of honeyeaters often occur in one area, usually with one to three species of lorikeets and the Silvereye. How are they able to coexist? How important is interspecific competition?

Honeyeaters depend both on insects and nectar or some alternative source of simple carbohydrates. Different species differ in their dependence on these two main food sources (Pyke 1980). The shorter-beaked species (*Meliphaga, Lichenostomus, Melithreptus, Manorina*) tend to take more insects and also manna, lerps and honeydew. Longer-beaked species are on the whole more dependent on nectar as a source of energy (*Phylidonyris, Acanthorhynchus, Anthochaera*). There are of course great differences between seasons and sites. Similar species in the short-beaked group often occupy different habitats, as discussed in Chapter 3.

Among the nectarivorous guild there are differences in the types of flowers visited by different species. This is rarely because species have beaks shaped to fit particular flowers, as is found in the hummingbirds (e.g., Snow and Snow 1972), although the long-billed Eastern Spinebill (*Acanthorhynchus tenuirostris*) commonly visits long tubular flowers. On the whole the larger species tend to monopolize the richer, more reliable nectar sources. These they may defend, and so relegate smaller species to more dispersed or poorer nectar sources (Figure 21; Ford 1979; Ford and Paton 1982). A similar hierarchy of dominant to subordinate species is

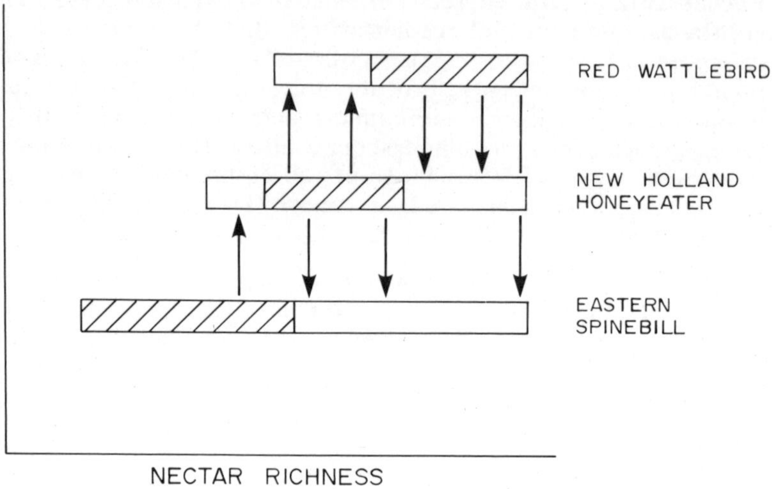

Fig. 21. Coexistence of honeyeater species along a continuum of nectar richness. Potential niches indicated by rectangles, actual niches by hatched areas. Vertical arrows indicate aggressive exclusion (downwards) and restriction by exploitation (upwards).

also found in assemblages of hummingbirds (Feinsinger 1976). Birds may even defend flowers against bees. This pattern of coexistence has only been found in a few sites so far, and many more guilds need to be investigated closely.

There is also evidence for a similar interspecific hierarchy along a gradient of resources amongst the short-beaked species. Miners are relatively large and aggressive (Dow 1977), and exclude smaller honeyeaters and other birds from localities where carbohydrates like lerp and honeydew are abundant (Wykes 1985). Yellow-tufted Honeyeaters (*Lichenostomus melanops*) are also localized and aggressive, whereas White-eared and Yellow-faced Honeyeaters (*L. leucotis* and *L. chrysops*) tend to be forced into poorer areas or to occur where there are undefended rich sources (Wykes 1985).

Interspecific Competition in Bird Assemblages

We are now in a better position to assess the role that interspecific competition plays in determining the composition and structure of assemblages of insectivorous and nectarivorous birds in Australia. Evidence for the importance of competition in communities has been gleaned from a wide variety of sources (Giller 1984). Until recently much of it has been indirect and inferential. If one compares the niches of coexisting species one almost invariably finds differences in food, foraging site or time of activity. If resources were limiting then these differences would reduce interspecific competition. Alternatively they may simply reflect the different morphologies and histories of the species. Sometimes pairs of species are more different ecologically where they coexist than where they occur separately (e.g., Diamond 1986). Again, this is consistent with competition but is not strong support for it. One example, that of the small and medium ground finches, *Geospiza fuliginosa* and *G. fortis*, on the Galapagos Islands, has been investigated in great detail (Grant 1986). On the small island of Daphne Major *G. fortis* has a smaller beak on average than it does on those islands where its smaller congener also occurs. Small individuals can survive food shortages on Daphne Major by consuming small seeds that are taken elsewhere by *G. fuliginosa*.

Frequently one finds species that are very similar ecologically and whose ranges abut but do not overlap. Competition might be preventing coexistence in such cases, but there could also be unmeasured differences between the habitats and ecological requirements of the two species. Finally, sometimes there are changes in the location of boundaries between pairs of species over moderate periods of time. All of these sources of evidence can be found amongst Australian birds and together provide a reasonably strong case for competition. However, they are open to alternative interpretations, such as predation or the influence of less obvious environmental factors.

Interspecific competition can take place in two major ways: by interference and by exploitation. *Interference* is the process by which species prevent each other from obtaining critical resources by territorial behaviour or other forms of aggression. In *exploitation*, critical resources for two or more species are simply depleted without direct interaction. If these processes can be observed or measured then competition is strongly implicated.

As has been previously discussed, nectar-feeding birds in Australia and on other continents are frequently aggressive and territorial towards other species. Wattlebirds (*Anthochaera*) and

New Holland Honeyeaters (*Phylidonyris novaehollandiae*) often exclude other honeyeaters from flowering banksias, eucalypts and other plants. This is clearly interference, though in one direction rather than mutual. Defence is rarely, perhaps never, totally successful. The subordinate species may sneak in, or join flocks and simply swamp a territory. In addition even their attempts to intrude drain the energy of the owner. There is much direct evidence that nectar sources are depleted during the day, often by quite early in the morning (Ford 1979; Ford and Paton 1982). Thus an important condition for interspecific competition, limiting resources, is at times satisfied. Therefore I maintain that there is strong interspecific competition in honeyeater assemblages, with the larger species tending to win the interference battles, but the smaller species, because they have lower energy requirements, being the more efficient exploiters. Despite this competition, co-existence can be permitted where the richness of nectar sources varies from place to place or through time. Rich patches can be defended by larger species and smaller species can exploit the poorer intervening areas. Probably similar patterns are shown where honeyeaters are feeding on other carbohydrate sources like lerp, honeydew and manna. These are harder to measure though.

Interspecific territoriality is less obvious among more strictly insectivorous birds (Murray 1981). In Australia Flame and Scarlet Robins (*Petroica phoenicea* and *P. multicolor*) defend territories against each other (Loyn 1980). It is also far harder to measure depletion of insects than of nectar, as the former are replenished over a longer time. Figure 16 indicates that birds can influence the abundance of insects, but the pilot study on which it is based needs to be repeated on a larger scale, and in other places.

Most ecologists believe that the most convincing test of interspecific competition comes from experiments where one species is removed from sympatry, or added to a second in allopatry. Changes that indicate competition are corresponding expansion or contraction of niches, altered breeding success or changes in population sizes. Schoener (1983) and Connell (1983) reviewed the surprisingly large number of experiments testing for competition. Most experiments provided evidence for interspecific competition, but not for all pairs of species, or at all sites or for all of the time. Removal experiments are difficult to perform on birds as they are mobile and can easily re-invade experimental areas. This has not prevented attempts to remove or at least suppress one or more bird species (e.g. Högstedt 1980; Minot 1981; Garcia 1983). These studies have invariably shown increased breeding success or expanded niches in the species relieved from competition. To my knowledge only a single experiment of this type has been carried out in Australia. This was the experimental removal of Bell Miners by Richard Loyn and his colleagues (1983) from a site in Victoria. Other honeyeaters, pardalotes and other insectivorous birds increased rapidly in the experimental site, but not in a control site where Miners remained. This provides clear evidence that the aggression of the Miners had excluded these other birds from an area potentially rich in insects. This experiment needs to be repeated, perhaps with Noisy Miners (*M. melanocephala*). Re-location may be preferable to removal by shooting. This would have the advantage that a species could be removed from one site and added to another community from which it was absent. Such an experiment would be most appropriate in highly sedentary species such as thornbills.

Whereas some experiments on birds have indicated that competition occurs others have not and it is unclear how frequently competition is operating. Weins (1977) suggested that for most of the time interspecific competition is not a major problem for birds, but that periodic resource shortages ("ecological crunches") elevate its importance. A pattern such as this seems likely in the more arid parts of Australia where good years alternate with bad.

In conclusion, there is strong evidence from a variety of sources that interspecific competition is important in determining the structure of nectarivorous bird assemblages in Australia. Convincing experiments that involve removal or addition of species would be very hard to perform on these very mobile birds. In insectivorous bird assemblages the evidence is inferential, but supported by two small-scale experiments indicating depletion of food and aggressive exclusion respectively. Interspecific competition could well operate by modifying the niches and abundance of species within communities, rather than by limiting which species can coexist. Features such as the presence of suitable foods, substrates and nesting sites are probably more important in determining the composition of avian assemblages, than are bird species that are already present. Perhaps a major exception is where sibling species (e.g., *Lichenostomus* honeyeaters) occur in neighbouring habitats.

Succession

Succession, the gradual replacement of one community by another, has principally been applied to the plant component.

Familiar examples include the progression from grassland, through shrubland to woodland, and from lake to marsh to woodland. As the vegetation changes so too do the species of animals dependent upon it. Fuller (1982) described the bird assemblages found at different successional stages in Britain. The early stages like downland, lowland heath and fen have few individuals, mostly of one or two species. At the opposite extreme mature deciduous woodland may have as many as 80 species of birds, none of which occurs in the first stage. Succession in bird assemblages as such does not appear to have been studied in Australia, at least in the long term. However, comparisons have been made between different stages of vegetational succession. Grassland assemblages are typically very simple with the single species of pipit (*Anthus novaeseelandiae*) and lark (*Mirafra javanica*), occurring along with quail (*Coturnix*). Other birds like parrots, pigeons and ibis feed in grassland but are dependent on trees for breeding and roosting. Birds of prey also hunt overhead. Australia has no counterpart to the rich communities of pipits, larks and sparrows found in North American or African grasslands (e.g., Wiens and Rotenberry 1980). Shrublands tend to be occupied by species that occur in the understorey of native woodland or forest, although in arid regions they may possess a distinct bird assemblage.

Although succession itself has not been followed throughout its course, the effects of natural or human events which reverse the successional sequence have been studied. The most important natural event in Australia is fire, the frequency of which has probably been altered by Aboriginal and European man. Cyclones and landslides can also devastate forests, as can rare events such as earthquakes and volcanic eruptions. Human activities that are important are logging and clearing for agriculture, which have been widespread, and mining, which is local in its effect. I shall discuss fire on bird assemblages here, but leave the effects of agriculture, logging and mining until Chapter 11.

Fire and bird assemblages of forests

Fires vary greatly in their intensity and consequently in their impact on the forest and its bird assemblage. Eucalypt forests recover dramatically from fire; from the apparent devastation trees produce epicormic shoots and shrubs sprout from lignotubers or germinate from seeds. Grasses and herbs may become abundant for some months afterwards. The release of nutrients in the ash may encourage very rapid growth of nutritious tissue. This may lead to a build-up in leaf-eating insects. Also, many plants may flower and set seed prolifically within a year of fire. Therefore fires, unless they are excessively severe, do not take the ecosystem back to its earliest successional stages.

Perhaps surprisingly, individual birds can survive fires (Recher, Allen and Gowing 1985), although mortality may be high (Wegener 1984). With mild or moderate fires the density and diversity of birds may actually increase for 2–3 years afterwards (Christensen *et al.* 1985). Seed-eating birds like doves, pigeons and quail take advantage of the easily available seeds straight after a fire (Recher, Allen and Gowing 1985) and firetails may be common as grass sets seed. Lorikeets, honeyeaters, silvereyes and pardalotes, all feeders on carbohydrates, often become common as plants flower and new growth becomes infested with psyllids and other insects. Shrub-dwelling species like scrub-wrens and Brown Thornbills (*Acanthiza pusilla*) may decline initially (Christensen *et al.* 1985), but increase later (Recher, Allen and Gowing 1985). Canopy-dwelling insectivores may be affected little by ground fires but may decline after severe fires that burn or scorch the crowns of trees. The Golden Whistler (*Pachycephala pectoralis*) may be replaced by the Rufous Whistler (*P. rufiventris*), its congener that favours more open habitats. Robins which pounce on prey on the ground are not seriously affected by fire (Marchant 1985). Other species that favour more open habitats like Dusky Woodswallows (*Artamus cyanopterus*), Magpies (*Gymnorhina tibicen*) and Tree Martins (*Cecropis nigricans*) may increase until the canopy becomes closed again.

Thus there are changes in the bird assemblage following fire, whether it be wildfire or controlled or accidental burning. These changes are not dramatic and except after severe or extensive fires the bird assemblage will have recovered its original composition within a year or two of the fire.

Wetland Succession

Lakes, especially shallow ones, gradually silt up and develop marshy or intermittently moist fringes. These extend outwards until little open water remains. In time shrubs and trees colonize and wet forest results. Again this succession has not been followed by ornithologists. However, Jon Fjeldså (1985) undertook a detailed comparison of the waterbird assemblages of southeastern Australia. To some extent his classification into smaller groupings reflects different stages of the succession from lake to dry land. Other factors like soil type and its effect on emergent or other vegetation and permanence of the water body greatly influence the bird assemblage. Also, single water bodies may have highly dynamic bird assemblages, depending not only on water level but also on what is happening in nearby or more distant areas (Briggs 1977).

In Table 16 I have arranged Fjeldså's assemblages in a sequence and have listed typical species in each assemblage. Most of these species occur in other assemblages and some species, e.g., Grey Teal (*Anas gibberifrons*), may occur in all types of water body.

Australia naturally lacks deep infertile (oligotrophic) lakes such as those that are common in northern Europe and the cooler parts of North America. Some reservoirs and water storages in eastern Australia can be deep. Howarth and Grant (1982) found few species and individuals on deep storage dams near Wollongong. They were mostly Coots (*Fulica atra)*, small grebes and cormorants. There are no counterparts to the fish-eating divers (loons) and mergansers that favour these habitats in the Northern Hemisphere, though the Great-crested Grebe (*Podiceps cristatus)*, occupies the deep lakes of South Island, New Zealand (Westerkov 1971).

Australian swamps are diverse and have quite similar bird assemblages to those of Scandinavia and the Andes which Fjeldså also studied.

Stability

Stability is a poorly understood, indeed misunderstood, concept in ecology. One of the main reasons for this is that stability has been used to mean a wide variety of phenomena. Pimm (1984) has provided alternative names for some of these phenomena. For a start it has been used to describe environmental variation in climate, as well as to describe the consistency of composition, relative abundance of species and ability to recover from disturbance in the biotic community (e.g., Ricklefs 1980a; Giller 1984). In addition, it has been used to describe low levels of changes over very short to very long periods of time, some of which are oscillating and predictable, whereas others are progressive.

In the very short term there are changes in temperature and humidity during the day. Equatorial forests change little and provide a more gentle environment for their animals and plants whereas deserts and high mountains fluctuate more extremely.

Table 16. Wetland habitats in southeastern Australia, arranged in an approximate successional sequence, with typical species or groups of birds (adapted from Fjeldså 1985).

Deep lakes/swamps		Permanent swamps with		
stable	unstable	emergent or floating plants	Temporary swamps	Open shoreline
Great-crested Grebe	Darter	Black Duck	Hoary-headed Grebe	Black-winged Stilt
Hoary-headed Grebe	Pied Cormorant	Swamphen	Glossy Ibis	Sandpipers
Hardhead	Great Cormorant	Marsh Harrier	Straw-necked Ibis	Shelduck
Musk Duck	Pelican	Little Pied Cormorant	Pink-eared Duck	Maned Duck
Blue-billed Duck		Little Black Cormorant	Black-tailed Native-hen	Plovers
Whiskered Tern		Sacred Ibis	Red-kneed Plover	
Black Swan		Egrets		
Coot		Herons		
		Bitterns		
		Spoonbills		
		Dusky Moorhen		
		Reed Warbler		
		Jacana		
		Rails		
		Crakes		

There are also day-to-day changes in weather, which can alter the availability of and requirement for food. Seasonal changes are a feature of nearly all environments. Growth of plants may be slowed or prevented by low temperatures in temperate regions or a shortage of moisture in mediterranean and subtropical environments. Consequently animals may suffer relative food shortages for several months of the year. There are also marked changes between years in such factors as rainfall. This tends to be most extreme in drier habitats, and semi-arid regions alternately experience years of drought and of abundant growth.

In the much longer term one habitat can replace another. Partly this is due to successional change, when, for instance, rainforest gradually spreads into wet eucalypt forest. Some changes though are due to long term environmental change, as in the wetter and drier periods of the Pleistocene, when the forests expanded at the expense of the deserts and then contracted again.

Most environments experience freak and unpredictable events, though these may in fact not be so unusual (Weatherhead 1986). These may be climatic conditions outside the normal range, like droughts in normally mesic areas, or devastating events like fire and cyclones. Where these events are relatively frequent, ecosystems may return quickly to their original structure and composition. This may not be the case where they are very infrequent. In the previous section I described the ability of eucalypt forest to recover rapidly from fire. Rainforests though may take decades to do so.

With stability meaning so many different things in ecology, it may be sensible to use other terms whenever possible. *Seasonal*, and its converse *aseasonal*, should be used to describe changes within the year. The term *equable* could be used to describe environments with low daily or annual ranges of conditions. *Predictability* perhaps best describes day-to-day and year-to-year changes. Successional change can often be used to describe longer term changes. Communities that change little over long time periods can be described as *persistent*. Indeed the term stability need not be used to describe abiotic changes but can be restricted to progressive or fluctuating biotic changes; in the composition of communities, or in population densities of constituent species. *Resilience* could be used for the ability to recover from severe disturbances. *Resistance* is a measure of the ability to withstand change after a perturbation (Pimm 1984).

I shall deal with stability, in its broadest sense, in various places in this book. In Chapter 9 I shall discuss the importance of seasonal changes for migration and the influence of year-to-year predictability on nomadic movements of birds. I shall discuss how both seasonality and unpredictability may contribute to the evolution of co-operative breeding in Australian birds in Chapter 8, and how they affect breeding seasons and effort in Chapter 5. The resistance and resilience of bird assemblages in the face of human activities such as logging, agriculture and mining will be discussed in Chapter 11.

I propose to discuss here only the effect of day-to-day changes and changes over several years on the composition of bird assemblages. There is in fact very little information on either aspect.

Day-to-day Changes

Most of southern Australia experiences marked changes in weather from day-to-day. These are perhaps most extreme in spring when days with warm dry northerly winds alternate with days when cool air and rain is brought in from the southern ocean. Energy requirements of birds are higher on the cooler days and the abundance of food such as insects and nectar may be lower. Many insectivorous species start breeding in early spring but characteristically stop with cool weather then restart as temperatures rise again. Such fluctuations may prevent particular species from establishing themselves in a community or could be responsible for marked annual fluctuations in population size. These aspects need to be studied. In the case of nectarivorous birds I know of one case where day-to-day unpredictability probably affects the structure of the bird assemblage. New England National Park has flowering banksias which *on average* are highly productive. In most places where banksias are abundant, medium-sized and large honeyeaters are common (McFarland 1986a). However, New England National Park experiences cold nights in winter and nectar production is closely correlated with minimum temperature (McFarland 1985). Days of abundant nectar alternate with days of almost no nectar following mild and cold nights respectively. The larger species may have difficulty in maintaining their energy reserves and the small Eastern Spinebill (*Acanthorhynchus tenuirostris*) is the most common honeyeater. It is able to exist on poorer nectar sources than the larger species and survives lean times by storing fat from more productive times and perhaps by reducing metabolic rate. During extended cold spells though, even Spinebills have to leave the area (Ford, Chan and Ambrose, unpublished).

Longer-term Persistence of Bird Assemblages

There appear to have been no detailed and methodical studies of bird assemblages over more than a five-year period in Australia, which are comparable with those carried out in Europe (e.g., Solonen 1986). However, several areas have had frequent visits by ornithologists who have noted the presence and absence of species at different times. I have chosen to compare three areas that differ in rainfall and habitat. Michael Brooker and his colleagues noted birds in the northwestern Nullarbor Plain over 10 years (Brooker *et al.* 1979). Mary Whitmore and her colleagues recorded birds in brigalow country near Meandarra in southeast Queensland for five years (Whitmore *et al.* 1983). I have added information from Eastwood State Forest near Armidale, where a succession of students, assistants and I have worked since 1979. Species were separated into (1) residents, present all year and every year, (2) summer and winter visitors, recorded each year in the appropriate season, (3) irregular visitors, present in some years, often for some time, but not in others, and (4) vagrants, recorded only a few times, usually for only a day or two. The numbers of each species in each category for each area are presented in Table 17. As average rainfall increases so does the proportion of regular species (residents and migrants), from 40% on the Nullarbor; to 56% at Meandarra and 64% near Armidale. The irregular species form a larger component of the avifauna on the Nullarbor; indeed Brooker *et al.* (1979) had a separate category of nomads, comprising 15 species. Whitmore *et al.* (1983) did not specify nomads, but I would class only seven species at Armidale as nomads. Most of the irregular species there were passage migrants occurring in spring or autumn. All three studies covered periods of extreme drought. On the Nullarbor many species were absent or very scarce during the drought and this is true of some species at Meandarra. During the drought Armidale received several species that are typically found further inland — e.g., Red-capped Robin (*Petroica goodenovii*), Plum-headed Finch (*Aidemosyne modesta*) and White-browed Woodswallow (*Artamus superciliosus*).

As well as species that were periodically absent, Brooker *et al.* recorded two species that apparently have disappeared over the last 40 years, the Chiming Wedgebill (*Psophodes occidentalis*) and the Redthroat (*Sericornis brunneus*). Galahs (*Cacatua

Table 17. An indication of the stability of the landbird assemblage in three sites. The category of irregular visitor includes nomads in Brooker *et al.* and transients in Whitmore *et al.*

	Nullarbor	Meandarra	Armidale
Habitat	Acacia woodland/shrub steppe	Brigalow	Eucalypt woodland
Source	Brooker *et al.* (1979a)	Whitmore *et al.* (1983)	Ford *et al.* unpubl.
Rainfall	c 200 mm	585 mm	790 mm
Total number of species	78	137	98
Resident	30	60	43
Regular summer visitor	1	14	18
Regular winter visitor	0	3	2
Irregular visitor	41	29	23
Vagrant	6	31	12

roseicapilla) had colonized Brooker's area, as well as Armidale, and Crested Pigeons (*Ocyphaps lophotes*) were apparently doing so. These changes are at least partly attributable to changes caused to the habitat by human activities, from grazing by stock and rabbits and through the provision of water.

Drier habitats, which have less predictable rainfall, have bird assemblages whose composition changes more than those of wetter, more predictable habitats. However, this conclusion is tentative as different observers may classify their species differently and differ in their pattern of visits to their sites. A more thorough comparison using all sites where observations have been made over five or more years seems merited. It is also clear that a standardized monitoring of bird assemblages over a long period should be started as soon as possible throughout Australia. Patterns of change should be related to climatic and other environmental changes.

The question of stability has featured in one other controversy in community ecology: whether more diverse communities are more stable. Elton (1958) proposed that more diverse communities were better able to buffer environmental fluctuations than were simple ones. This was an intuitively attractive concept as it helped to explain why outbreaks of pests were more common in crops and plantations of monocultures than in complex forests.

However, it appears that theoretically there is no reason why complex communities should be inherently more stable, indeed they could be less stable than simple ones (May 1972). Part of the reason for this misunderstanding was that again stability, and indeed complexity and diversity, were used in several ways (Pimm 1984). It appears that diverse communities may indeed be more persistent from the point of view of population fluctuations. However, this is probably because environmental stability can lead to both more species coexisting and also to their population levels fluctuating less than in more unpredictable environments. As mentioned in Chapter 1, bird assemblages of eucalypt forests are somewhat more diverse than those of rainforests and much more diverse than those of deserts. The reason for the former is that the area of rainforests in Australia is small and historically has been even smaller. The reason for the latter is probably that deserts are much less predictable environments than eucalypt forests. It will be a long time before there are sufficient data on population fluctuations of birds from these three habitats to permit us to relate seasonality and predictability of the environment to the persistence, resilience and resistance of the bird assemblages.

Summary

Birds are important components of communities. In forests they play a part in controlling herbivorous insect populations. In Australia they are major pollinators of many sclerophyllous plants, dispersers of rainforest plants and consumers of seeds in most habitats.

The diversity of a bird assemblage depends on the structural diversity of its habitat, the productivity of the area and its proximity to similar environments. Bird assemblages in forests and woodlands are separated into guilds that forage on different foods or from different substrates. Guilds may be subdivided by foraging method and height and by the type and richness of food. The presence of appropriate foods, foraging substrates and other features determines which species can be present in a community. Interspecific competition may determine which species actually are present or more often what niche these species occupy. The evidence for competition is stronger for nectarivorous birds than for insectivorous birds. Honeyeaters may deplete their food on a daily basis and aggressively exclude other species.

Communities undergo succession, for instance from grassland or open water to woodland. Bird assemblages change in composition and diversity along these gradients. Fires to some degree take forests to earlier successional stages. Eucalypt forests, and the bird assemblages inhabiting them, are particularly resilient to fires, unless they are very intense.

Stability has been used in a variety of ways in ecology: seasonal changes in climate and food availability; predictability of year-to-year changes; changes in population density and composition of communities over moderate to long periods and the resilience of a community to disturbance. Whenever possible more specific terms should be used. It appears that bird assemblages change more between years in the drier and less predictable habitats in Australia, though information is scanty.

5. Breeding

There is intense selection on any organism to maximize its lifetime reproductive success. This necessitates a compromise between devoting energy and time to enhancing adult survival and devoting them to the production and care of offspring. Different species partition their effort between survival and reproduction in different ways. At one extreme, long-lived species may strive to occupy favourable habitat and avoid predators while only making modest reproductive attempts, whereas other species grow rapidly, put much effort into reproducing and usually do not live long. The former are often called *"K"-selected* species, the latter *"r"-selected* ones. Most species lie between the two extremes. Generally "K"-selected species tend to be larger than "r"-selected ones, they tend to live in environments that do not experience physical extremes or food shortages, but they often experience stronger competition, especially from conspecifics (Table 18; Pianka 1974).

Most insects tend to be nearer the "r"-selected end of the spectrum whereas vertebrates, including birds, are more "K"-selected. Among birds, small passerines, specially those in cold climates, tend to be more "r"-selected than are raptors or seabirds.

Perhaps the way to illustrate how evolution has influenced reproductive strategies is to pose a series of questions and compare the ways in which different species have responded. The questions are:

1. At what age should breeding commence?
2. When is the best season to breed?
3. How many breeding attempts should be made per season?
4. How many eggs should be laid per clutch?
5. How much parental care should be offered to the offspring?

Obviously these questions are related. Individuals that start breeding when they are young may lay smaller clutches.

Table 18. Characteristics often displayed by "r" and "K" selected species. Note that most species lie between these extremes and that all characteristics of one strategy are rarely combined together (from Pianka 1974).

	"r"-selected	"K"-selected
Climate	Seasonal and/or Unpredictible	Equable
Population	Fluctuating	Stable
Competition	Rarely strong	Often strong
Growth	Rapid	Slow
Body Size	Small	Large
Start of breeding	Early	Late
Reproductive effort	Single major attempt Large clutches	Repeated minor attempts Small clutches
Motality	Density independent*	Density dependent*
Length of life	Short	Long
Selection maximizes	Reproductive Rate	Competitive Ability

*These terms will be defined in Chapter 8.

Individuals that care for their offspring for a long period may forgo a second breeding attempt in a season. However, initially I shall consider them independently.

Delayed Breeding

Most small passerines at least attempt to breed when they are about one year old. The Zebra Finch (*Poephila guttata*) and Budgerigar (*Melopsittacus undulatus*) are sexually mature in captivity at about three months after hatching (Marshall and Serventy 1958), whereas the Golden-headed Cisticola (*Cisticola exilis*) has been recorded breeding in Japan 27 days after fledging (Ueda 1985). Larger species, and in particular seabirds may delay their first breeding attempt for two or more years. There are two main reasons why breeding is delayed. Individuals may delay breeding until they have developed the skills necessary for a reasonable chance of success. Such species are long-lived. Secondly, individuals may be prevented from breeding, even though they would be capable of doing so, because more dominant individuals have monopolized all the suitable habitats or nest sites, or because mates are scarce.

Many seabirds appear to delay breeding until they have developed specialized feeding skills; they frequently need to travel great distances from feeding grounds to feed their chicks. There have been many classic and intensive studies of the breeding biology of Australasian seabirds, which have contributed importantly to the development of theory relating to breeding ecology. Richdale (1957) studied the Yellow-eyed Penguin (*Megadyptes antipodes*) near Dunedin, and found no birds breeding in their first year (Plate 40). Eight per cent of males and 48% of females bred in their second year. Almost all females bred by their fourth year but nearly a third of males still did not. With an adult annual survival of 87% there is perhaps no urgency to start breeding.

Dom Serventy and his colleagues have studied Mutton-birds (Short-tailed Shearwater — *Puffinus tenuirostris*) in the Bass Strait for over 30 years. Mutton-birds do not start breeding until 5 to 8 years old (the females on average younger than the males). As the annual survival of a breeding adult is over 90%, most birds can expect 10 or more breeding attempts and some certainly achieve 30. Probably the reasons why breeding is delayed so long is that Mutton-birds experience a very demanding annual cycle, including a circum-Pacific migration (see Chapter 9); they lay an egg that weighs one sixth of their body weight and may travel great distances to collect food for their nestlings (Serventy 1963, 1967).

A variety of species have a significant proportion of individuals that do not breed until after their first year because they lack the opportunity to do so. These include many co-operative breeders such as fairy-wrens and babblers and also polygamous species such as bower-birds. These will be discussed in more detail in the next chapter.

In many species that do breed in their first year, young birds tend to start breeding later in the season, perhaps because their lack of experience means that they take longer to build up sufficient energy reserves for egg-laying and incubation. Establishment of a territory, the attraction of a mate and nest-building may also take first-year birds longer than experienced breeders.

Breeding Seasons

In the temperate parts of the Northern Hemisphere birds traditionally breed in spring and early summer (April to July). This is the time of moderate temperatures and long days. As one moves towards the equator breeding starts earlier and also the season extends over a longer period (Murton and Westwood 1977). Temperatures are mild for a longer period in the tropics and day-length does not vary much over the year.

However, the major *ultimate factor* in determining the breeding season is food supply. It is advantageous for birds to breed at such a time that their young are in the nest or are recently fledged when there is a peak in abundance of food (Lack 1954). Insects tend to be most abundant in deciduous forest when, or shortly after, the new leaves emerge in late spring. Later, the leaves become tougher and contain more harmful chemicals so that caterpillars in particular decline. Many flying insects continue to be common throughout summer. That food is a major factor is shown by the differences between related species in their breeding season when they rely on different foods. For instance, among European finches those that feed their young on caterpillars breed earlier than those that eat seeds (Newton 1972).

In birds of prey the European Kestrel (*Falco tinnunculus*) breeds early, with the production of young coinciding with the spring peak in small mammals, whereas the bird-eating Sparrowhawk (*Accipiter nisus*) breeds in early summer when fledglings are abundant. The Hobby (*F. subbuteo*) breeds even later and feeds on young dragonflies and swallows, which are on the wing in late summer (Newton 1979). Some species breed outside the usual breeding seasons for most birds. Crossbills (*Loxia*) tend to breed from autumn to early spring and depend on conifer seeds, which are most available in winter (Newton 1972). Eleanora's Falcon (*F. eleanorae*), of the eastern Mediterranean, breeds in autumn, thereby exploiting the southward migration of young, naive passerines (Newton 1979).

Sometimes the association between abundant food and production of young birds is very close, as in titmice and leaf-eating caterpillars (Gibb 1950). These coincidences raise questions about the *proximate* (immediate) cues that birds use in starting to breed. Nest-building, egg-formation, egg-laying and incubation take at least one month, even in small passerines. In addition the gonads need to enlarge and in most species territories have to be staked out before breeding can begin. Therefore birds need to anticipate the period of food-abundance often by a couple of months. The most important initiator of breeding at high latitudes is day-length, principally an increase in the hours of daylight (Lofts and Murton 1968). This triggers the physiological and behavioural changes that are required before breeding can begin, i.e., enlarged testes and territorial singing. Though increased day-length may bring birds into breeding condition, the timing of breeding can be modified by the weather. Warm weather in spring may initiate a bout of nest-building and courting whereas a return to winter weather stalls any further breeding effort. Later springs consequently lead to both late caterpillar peaks and late broods of young birds (Perrins 1979). Birds also tend to breed later in poorer quality habitat, where food is less available.

In considering the timing of breeding, several other factors need to be taken into account. The behaviour associated with the early stages of breeding may be energetically demanding. Also, physiological changes, in particular egg-formation, require energy and especially protein. Further, birds that incubate their eggs need to start with a reserve of fat or be able to feed efficiently during the brief periods that they are away from the nest. In many species the male feeds the female while she is forming eggs or incubating. Food shortage early in the season may delay breeding so much that birds may miss the peak in abundance of their food. Addition of food may advance the breeding in some species. For instance by feeding Song Sparrows (*Melospiza melodia*) in winter Smith *et al.* (1980) were able to bring forward their breeding season by 3–4 weeks.

Birds that breed earlier than average will give their young a longer period of time in which to learn how to feed themselves before food becomes scarce. Young Great Tits (*Parus major*) from early broods are far more likely to be recaptured in subsequent years than are those from late broods, although sometimes early broods suffer from a late cold snap (Perrins 1979). Early breeding may give the adults other advantages; they are more likely to succeed in a second breeding attempt if their first fails or even to rear a second brood of young.

The number of breeding attempts per season will be closely related to the length of the breeding cycle and how long breeding is possible during the year. On the whole larger birds take longer to prepare for breeding and have long incubation, nestling and dependency periods. Rarely can they attempt a second clutch after successfully rearing one brood. Many small species though can make two, three or even more breeding attempts in a year. The advisability of making repeated attempts, particularly after successful breeding, depends partly on the period over which food is abundant (Figure 22). The caterpillars on which many small European passerines depend in deciduous forest are abundant for only a few weeks. To rear two broods, either the second one would have to be in the nest long after the peak in food, or the first and second broods would need to fall either side of the peak. Second broods are therefore rare after a successful first brood, at least in Britain (Perrins 1979). They may be more common in coniferous forests where caterpillars tend to be abundant for a

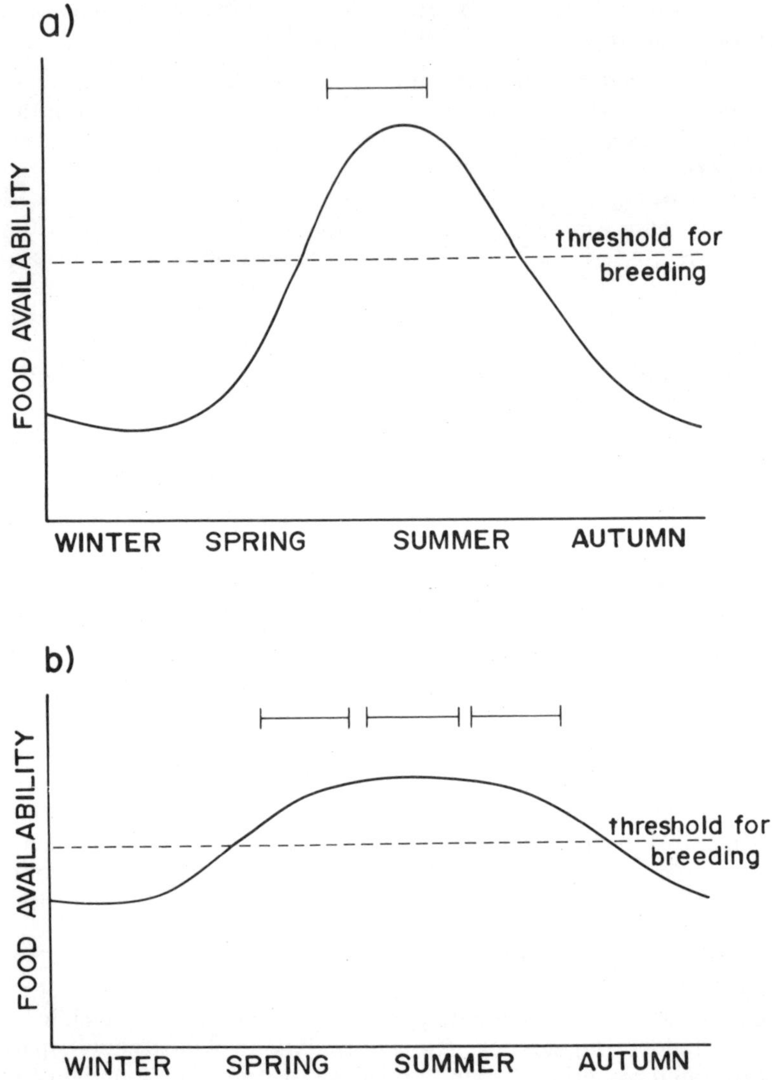

Fig. 22. Relationship between availability of food, breeding season, breeding effort, and number of attempts (horizontal bars) in two types of environment. a) is highly seasonal, with a period of shortage (usually winter) and a short period of great abundance. b) is less seasonal with no period of extreme shortage or abundance. Species occupying environment a) are more "r"-selected showing higher adult mortality, shorter breeding seasons with fewer attempts, but larger clutches than the more "K"-selected species in b).

longer time. More generalized insectivores than those relying on caterpillars or species taking plant and animal foods can fit two or three successful breeding attempts into a season, given a degree of good fortune.

Breeding, however, should not be considered alone in the annual cycle. All birds need to moult, at least once per year. New feathers require protein and energy, and birds with worn and partly grown feathers may be inefficient at flying and feeding. Consequently, breeding and moult rarely coincide, so that moult typically follows breeding in late summer and autumn when food is still rather plentiful and days relatively long. Many species also migrate in autumn and most, but not all, of these moult before leaving for their wintering grounds. Most also store fat which fuels their long-distance flights, or in sedentary species serves as a reserve for winter.

Whereas the timing and duration of breeding is well understood in north temperate birds, particularly in Europe, the patterns are much less clear in subtropical and tropical regions and in the Southern Hemisphere. Seasonal changes in daylength are small near the equator and temperature is rarely a limiting factor, though in arid areas it may be too hot to seek food for part of the day. Food is clearly important again, but it is rainfall that has the major influence on it. Where there is a well-defined wet season most birds breed during, just before or after it, when insects, flowers and fruits presumably are most abundant. Seed-eaters are often an exception as most seeds germinate at the beginning of the wet season, and thereby become unavailable. Seeds may be most abundant in the dry season. In equatorial forests rain may be plentiful year-round, but even there breeding is seasonal (Fogden 1972). Arid regions not only experience low rainfall but this may vary in amount and timing from year to year. Also, deserts may be cold at night in winter, which may reduce food and increase energy demands. Consequently birds may be more opportunistic in when they breed.

Breeding Seasons in Australasian Birds

The proximate and ultimate factors that control the breeding seasons of Australian birds are poorly known. There has been little field or laboratory experimentation on native birds and the spread in latitude and rainfall within the country means that breeding seasons and likely initiating factors probably vary from place to place.

By far the most comprehensive review of breeding seasons is that of Henry Nix (1976), who related annual cycles of birds to seasonal changes in plant growth. Growth of plants can be limited by extreme temperatures and lack of light or moisture. Light is perhaps rarely a problem but low temperatures reduce plant growth in winter and high temperatures are a factor in summer in the inland. Moisture is the major limiting factor to plant growth in most areas, in summer in the southwest and southern Australia (Mediterranean climates) and in the northern half of the continent in winter (monsoonal climate). Moisture is deficient in the inland most of the time, but is sufficient for most of the year along the eastern coast and almost throughout New Guinea. Nix predicted the best times of year for plant growth in different areas. These are spring in the southwest and south, and summer in the tropics and along the east coast and adjacent ranges. Inland areas may experience rain, and growth of plants (as long as temperatures are not too low), in almost any season, but particularly in summer from tropical depressions that result from cyclones. These do not occur every year though. In addition, autumn may be a good season for plant growth in the north and east and sometimes the southwest.

Nix (1976) predicted that birds should breed mostly during the season of highest plant growth and he then used data from specific sites to test his prediction. His predictions are generally supported (Figure 23). There are some anomalies, though, chiefly in areas where the climate is not markedly seasonal. In northeast Queensland and in the tablelands of New South Wales the main breeding season is somewhat earlier than expected. It is possible that heavy rain during the tropical wet season may make feeding difficult or destroy nests so that it is best to breed earlier there. Harry Bell has also suggested that thornbills and other small insectivores may start to breed early in New England thereby avoiding predatory reptiles and parasitic cuckoos, which are more prevalent in summer than in spring (Bell 1986b).

The pattern in New Guinea is interesting, as except in the southeast which has a marked winter dry season, plants grow well throughout the year. Despite this, such data as are available

---

Fig. 23 opposite. Relationship between plant growth (hatched) and main breeding seasons of different ecological groups of Australian birds (F = frugivores, N = nectarivores, I = insectivores and S = seedeaters) in a) a monsoonal climate, b) a Mediterranean climate, c) a semi-arid climate and d) a cool temperate climate. Note the very similar breeding seasons in all areas, and tendency for seed-eaters to breed later than other groups. (Adapted from Nix 1976).

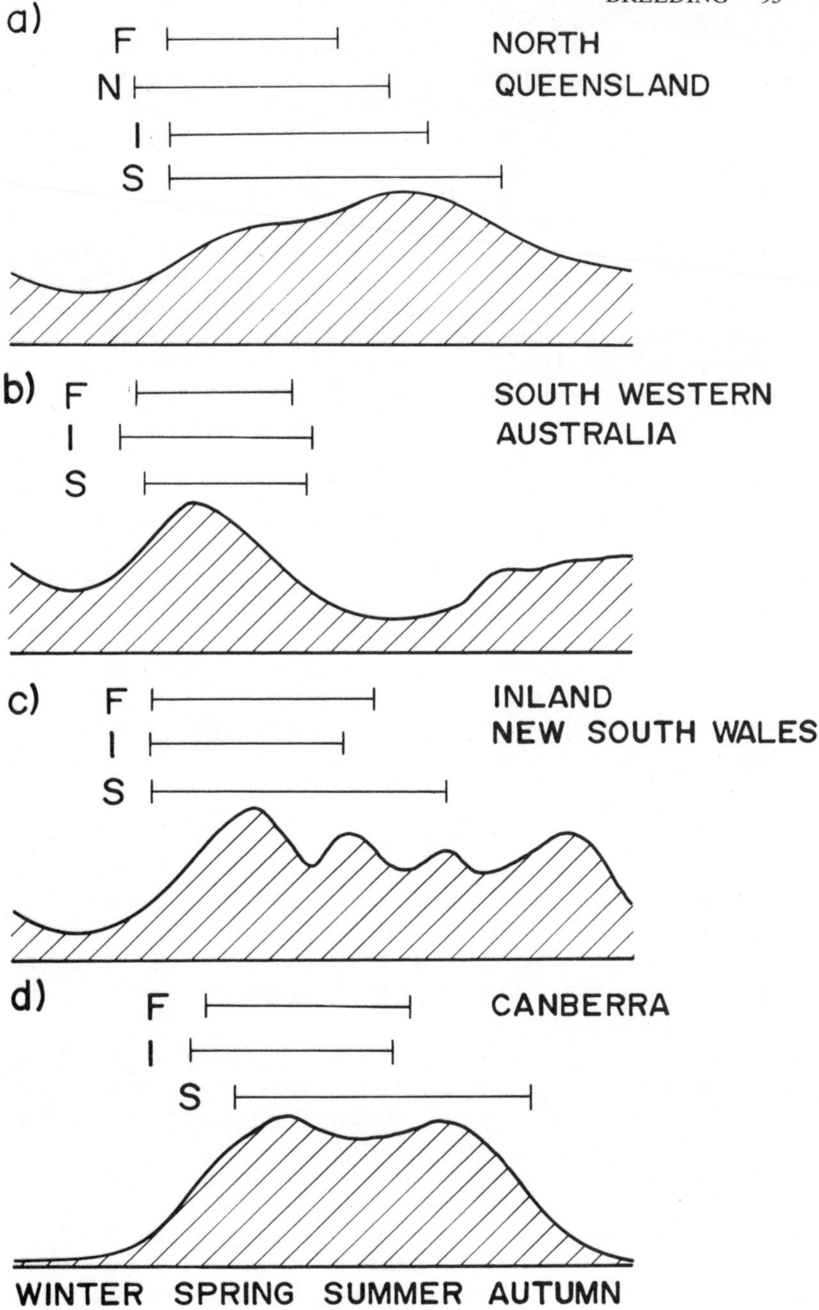
a)
F
N
I
S
NORTH
QUEENSLAND
b)
F
I
S
SOUTH WESTERN
AUSTRALIA
c)
F
I
S
INLAND
NEW SOUTH WALES
d)
F
I
S
CANBERRA
WINTER
SPRING
SUMMER
AUTUMN

indicate marked breeding seasons. Breeding follows the wettest period in coastal west Irian but precedes the wettest period in the highlands of west Irian (Nix 1976). Bell (1982c) collected an extensive range of data from rainforest near Port Moresby. Although nests were found in all months of the year there was a peak from September to December, preceding the wettest period again. A similar pattern was found in savanna woodland near Port Moresby (Bell 1982d). In contrast, birds in the New Hebrides (Vanuatu) and Flores show autumn peaks in breeding activity (Baker *et al.* 1940; Verheijen 1964). It appears then that birds tend not to breed in the dry season in the Australasian tropics, nor at the height of the wet season when plant growth is maximal. They may show peaks of breeding activity just before or just after the wet season. Perhaps each is equally suitable for most species, but there is an advantage in synchronizing the season, possibly because this reduces the risk of predation.

The breeding seasons of birds in inland Australia are poorly understood, as there is a scarcity of methodically collected data. Plant growth is less predictable there but tends to occur in summer and autumn in the northern half and spring in the south. This can lead to some species migrating north–south and breeding in the south in spring and in the north in autumn (Wyndham 1982). It is not known if the same individuals breed in both places and seasons. Rain tends to be unpredictable though and it is believed that some species of birds are opportunistic and breed whenever conditions are suitable (Serventy 1971). Zebra Finches are common birds in arid regions and display no apparent breeding season, but respond immediately to rain (Marshall and Serventy 1958). Davies (1977), in a longer term study in inland Western Australia, confirmed that rain could initiate breeding, but that temperature and food supply were also important influences. Zebra Finches bred regularly in spring, and did so also in autumn after summer rain. However, rain in autumn or winter did not initiate breeding.

Maclean (1976), in inland New South Wales, found a spring peak in breeding in a very wet year. There had apparently been little autumn or winter breeding despite good rain in autumn. This indicates that a spring breeding season may be normal, though Maclean suggested that after an extensive breeding season birds need to recover and opportunistic species will not breed for sustained periods even when conditions are favourable. Wyndham (1978), also found a spring breeding peak in the same wet year in northwestern New South Wales and southwestern Queensland.

Although spring and early summer is the main breeding season in southern Australia, in some years birds may also breed in autumn. Serventy and Marshall (1957) provided examples from inland southwestern Australia of many species breeding after late summer cyclones swept through the area. Some species, e.g., Red-tailed Black-Cockatoo (*Calyptorhynchus magnificus*), may breed regularly in autumn in this area (Saunders 1977). Honeyeaters may also breed in autumn in South Australia if the winter rains start early enough to initiate flowering of nectar-producing plants (Ford 1980), but breeding may be curtailed in mid-winter.

Throughout Australia, then, spring appears to be the main breeding period, though for different reasons in the north than in the south. Low temperatures prevent breeding in winter in the south and inland whereas low rainfall inhibits it in the north. Peak breeding may occur in summer in the coolest regions and perhaps where the wet season is late or not extensive. Maybe even in inland Australia daylength is the main factor that influences preparedness to breed, though rain may initiate actual breeding.

Nix (1976) discussed not only bird assemblages as a whole but also different ecological groups. He predicted that insects responded rapidly to new plant growth, and noted that flowers, fruits and seeds followed vegetative growth (Figure 23). Therefore within a community the sequence of breeding should be: insectivores, nectarivores, frugivores, then granivores. The data do not fully support these predictions although seed-eaters were generally late breeders. Fruit may take many months to mature and so may be available before the next burst of vegetative growth. Flowers too may develop in the previous season of vegetative growth to open early in the next season. This is clear in the Adelaide region where nectar is abundant in winter, whereas insects may be most available in spring. The more nectarivorous species of honeyeaters tend to breed earlier than those that are more dependent on insects (Ford 1980).

The waterbirds form a rather special group and many species are highly opportunistic. The Grey Teal (*Anas gibberifrons*) and Pink-eared Duck (*Malacorhynchus membranaceus*), in particular, may travel great distances to breed in areas where rain has recently fallen (Braithwaite 1976). When inland Australia has heavy rainfall these and other waterfowl, and cormorants, pelicans, terns and others may breed in vast numbers even in the normally dry salt lakes. As conditions dry out they depart for coastal regions.

There are few data on individual species, chiefly because the RAOU Nest Record Scheme is not yet extensive enough for analysis, except in the case of a few common species. Stephen Marchant has carried out several analyses, but even he is disappointed by the paucity of data on many aspects of breeding and the fact that most nest record cards are from one or two localities. To date the cards for Willie Wagtail (*Rhipidura leucophrys*), Welcome Swallow (*Hirundo neoxena*) and Yellow Robin (*Eopsaltria australis*) have been analysed (Marchant 1974, 1984; Marchant and Fullagar 1983). All have extensive latitudinal distributions and breeding seasons with a spring peak. Surprisingly there is little evidence that any of these species breeds earlier in the northern parts of their distribution. Welcome Swallows though do tend to breed earlier inland and at lower altitudes than in coastal or highland areas. Latitudinal uniformity may result from the differing limitations of temperature and moisture in south and north. However, further comparisons need to be made with more extensive sets of data.

Most of the previous discussion has addressed the timing of the start of breeding or the peak of the breeding season. I should now like to consider the length of the breeding season. At high latitudes conditions suitable for breeding are brief and hence most activity is squeezed into two or three months. Nearer the equator breeding is at least possible for most of the year. There is therefore a latitudinal gradient in the length of the breeding season, whether one considers the bird assemblage as a whole or only individual species. The longer breeding seasons at low latitudes may be because individuals take longer to breed, make more breeding attempts or the breeding season varies greatly among years.

Wyndham (1986) drew together data on breeding seasons from 35 locations around the world including 10 from the Australian region. He confirmed the increase in length of the season from temperate to tropical latitudes, for all sites as well as from Australian ones alone (Figure 24). Interestingly, the mean for any latitude in Australia and Africa combined (all Southern Hemisphere) was nearly two months longer than at the equivalent latitude in the other continents (mostly Northern Hemisphere). One reason for this anomaly might be that many Australian and African sites were in arid areas. Unpredictable rainfall might lead to opportunistic breeding, which, when data are added over many years, gives an apparently long breeding season. Wyndham therefore corrected the value for the length of the breeding season to remove the effects of latitude and tested it against rainfall. There

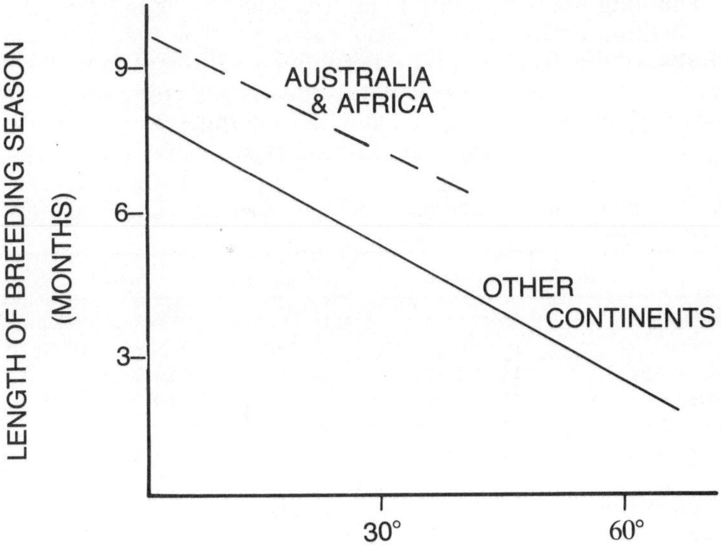

Fig. 24. Relationship between length of breeding season and latitude in Australia and Africa versus other continents (difference is significant; P<0.001). (From Wyndham 1986).

was no indication that birds in dry areas had longer breeding seasons than those from wet areas at the same latitude. Taken along with the marked seasonality of breeding in inland Australia discussed above, this does indicate that the opportunistic nature of breeding by birds in arid regions may have been exaggerated.

The reason why Australia, and perhaps the Southern Hemisphere generally should display long avian breeding seasons is unclear. Certainly aridity seems not to be the explanation. Possibly the greater area of the Southern Hemisphere that is covered by sea imposes a more equable temperature regime on the southern continents. Possibly for this, or some other reason, southern temperate regions have very few deciduous forests. Deciduous forests not only display highly seasonal patterns of abundance of food for birds but also leafless trees provide poor shelter for nests. As I have said before, long-term studies on bird behaviour and ecology in relation to the abundance of food, are required to test these suggestions.

The long breeding seasons of Australian birds indicate that individuals may make more breeding attempts than similar species in the Northern Hemisphere. Alternatively, different individuals may breed at different times of the species' breeding season or incubation, nestling or dependency periods may be particularly long in Australian birds. Such data as exist (Table 19) indicate

Table 19. Incubation and nestling periods of Australian birds with some European examples.

	Incubation	Nestling	Source
Thornbills	17–19 days	15–18 days	Woinarski 1985b
Pardalotes*	19–20 days	21–22 days	Woinarski 1985b
Yellow Robin	15–16 days	10–13 days	Marchant 1980
Whistlers	12–17 days	10–11 days	Marchant 1980
Honeyeaters	13–15 days	10–14 days	Marchant 1980
Fairy-Wrens	13–15 days	12–13 days	Rowley 1965
European Warblers	11–13 days	11–13 days	Woinarski 1985b
Titmice*	13–14 days	17–21 days	Perrins 1979
Thrushes	13–14 days	12–16 days	Simms 1978
Robin		14 days	Lack 1965

* hole-nesting.

that these separate stages of the breeding cycle are not exceptionally long in Australia though they are a few days longer than in comparable European species. In addition, repeated breeding attempts are common (Woinarski 1985b), possibly more so than on other continents. This is partly a result of the frequent failure of nests, but even after successfully rearing a brood, Australian birds frequently re-build and lay another clutch.

Clutch Size

In an area of science where quantification is difficult, the number of eggs in a clutch comes as a welcome relief. Extensive collections in the last century and field work and experiments in recent decades mean that we now have a good idea of the clutch size of many species, as well as a reasonable understanding of why it varies within and between species.

Probably the first comprehensive theory of clutch size was a remarkably simple one. Lack (1954) proposed that birds lay the number of eggs that will give them the largest number of independent offspring. Like all good hypotheses this was very testable and, with modification, it has stood the test of time reasonably well. Lack used data from Swifts (*Apus apus*) and Starlings (*Sturnus vulgaris*) to show that the most frequent clutch size was the one that led to the most offspring. Smaller clutches, even if successful were less productive, whereas in larger clutches there was less food per offspring and consequently the chance of starvation was greater.

By far the most extensive body of data has been collected on Great Tits and their relatives near Oxford and is summarized by Perrins (1979). Tits lay large clutches, typically 8–11 in most species, and as they nest in holes (usually nest boxes now) they experience low rates of predation. The rate of hatching is high, regardless of clutch size, so that Great Tit broods of seven to ten nestlings are the rule in England. Death from predation or starvation in the nest is quite rare so that larger broods generally lead to most fledglings. However, the parents have to work hard to bring caterpillars and other insects to large families, even though food is abundant. Consequently each individual in a large brood receives fewer feeds than one in a smaller brood, and indeed often smaller food items. This is partly compensated by the reduced needs for thermoregulation when surrounded by many other warm bodies. However, the result is that the weight at fledging declines greatly (Figure 25) with increasing brood size, especially above eight. Newly fledged young are difficult to follow, but they need to fend for themselves soon after leaving the nest, often when food is starting to decline. Heavier chicks appear to survive these critical weeks far better than light ones, as evidenced by recapture rates in the following year (Figure 25).

By combining the relation between brood size and weight at fledging, and the latter with survival rate, Perrins was able to calculate the most productive clutch sizes. These varied between years but tended to be close to, or slightly above, the most common clutch size (Figure 25). Lack's hypothesis is supported, but does need modification. What is most important is the number of young that subsequently breed. There is no value in producing many young to independence, if they survive no longer.

Some individuals appear to be able to lay larger than average clutches and rear many healthy young. Why is the average clutch size not still larger? Although clutch size is to some degree inherited it is also influenced by the age of the parent and the environment. Productive females are often experienced, or occupy particularly favourable habitats, or have hard-working mates.

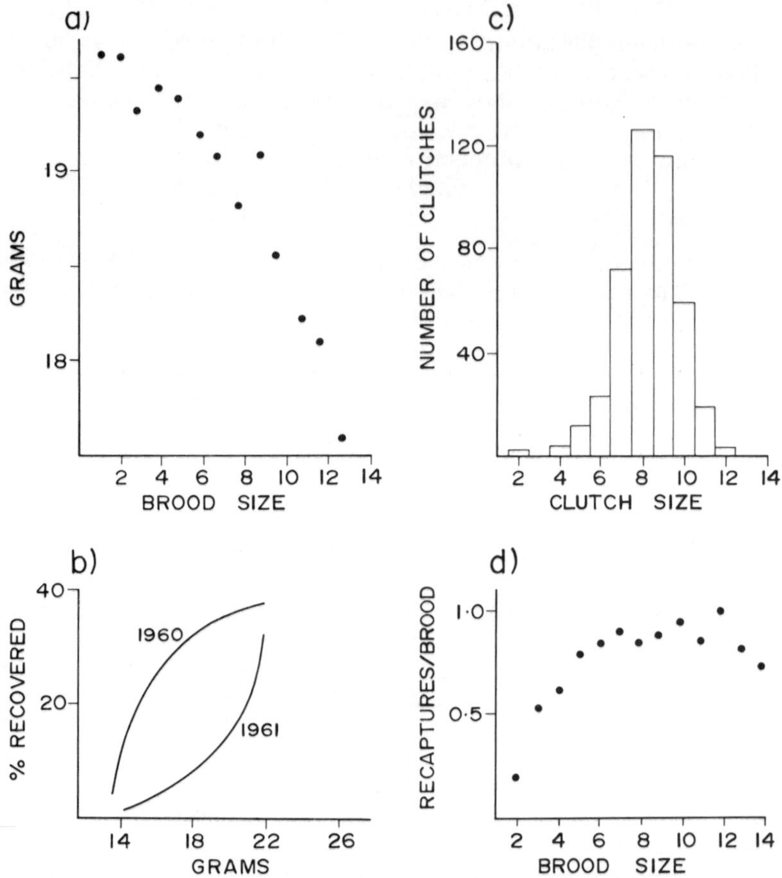

Fig. 25. Relationship of breeding success to clutch size in English Great Tits (from Perrins 1979). a) Weight of fledglings from different-sized broods, b) proportion of birds of different fledging weight that were seen again, c) frequency of different-sized clutches, d) estimate of number of birds surviving to next breeding season from different-sized broods.

Whereas a few birds are capable of rearing ten or more young, the majority may find that eight is the best that they can achieve.

A further complicating factor is that one breeding season should not be considered in isolation. Differences in conditions between years may mean that different clutch sizes are optimal. Natural selection favours those with the greatest productivity over their entire lifetime. Birds that over-extend themselves in one breeding season may find it hard to recover and survive the next winter, and hence breed again. Great Tits that reared twelve or more young were less likely to be breeding the next year than those that made more modest efforts.

Studies on the success of different clutch sizes, or the effect of adding extra eggs have not always shown that the most common clutch size is the best. Seabirds are apparently sometimes capable of rearing more young than they normally do (Nelson 1964). For instance Little Penguins (*Eudyptula minor*) can rear three young, yet they only normally lay two eggs (Dann 1988). It is possible that they are experiencing a particularly favourable food supply. More likely though, as they are so long-lived, they may jeopardize their survival chances, and future reproductive success, by working too hard in one season. The analyses of three Australian passerines (Marchant 1974, 1984; Marchant and Fullager 1983) indicated that larger clutches are more successful than the commonest clutch size (Table 20). There are no data on weights of fledglings or their subsequent survival though.

Table 20. Relative success of different clutch sizes in three Australian passerines.

	Welcome Swallow (Marchant & Fullager 1983)		Yellow Robin (Marchant 1984)		Willie Wagtail (Marchant 1974)	
Clutch Size	n	young reared/nest	n	young reared	n	young reared
1	4	0.25	3	0		
2	14	0.7	150	0.6	13	0.90
3	58	1.4	79	0.8	122	1.80
4	179	2.2			32	2.10
5	49	2.5				
6	1	6.0				

Within a species there may be differences between seasons or within a season. Superb Fairy-wrens (*Malurus cyaneus*) lay smaller clutches in dry years than in those with more rain. Many, but by no means all, species show a decline in clutch size through the season (e.g., Perrins 1979). This is probably because food becomes scarce for late broods. As yet there is no evidence of seasonal declines in clutch size in Australian birds (see Table 21); if anything it peaks in the middle of the season.

One frequent feature of clutch sizes within species is a latitudinal gradient. As one goes north in the Northern Hemisphere the number of eggs rises. For instance, in the European Robin

Table 21. Clutch size variation through the season in three Australian passerines (differences not significant).

	Willie Wagtail (Marchant 1974)	Yellow Robin (Marchant 1984)	Welcome Swallow (Marchant and Fullagar 1983)
Before October	3.05	2.16	1st clutch 3.95
October	3.16	2.34	2nd clutch 4.11
November	3.21	2.58	3rd clutch 3.87
After November	2.97	2.44	

(*Erithacus rubecula*) clutches in North Africa are typically four, whereas five is most common in England and six in Scandinavia (Lack 1965). Part of the reason for this is that days are longer in summer in the far north, allowing more time to collect food for the young. However, owls also show a latitudinal gradient in clutch size, and, being nocturnal, they have less time to feed in the north than in the south. It is more likely that food is simply more abundant in the breeding season at high latitudes, at least in relation to the density of breeding birds. Ashmole (1963) proposed that clutch size is correlated with the ratio of food at its most abundant to food at its scarcest. His theory was initially related to seabirds, but has been subsequently extended to land birds too (Cody 1966; Ricklefs 1980b). Highly seasonal environments do not only provide plenty of food in the breeding season, but also result in relatively low densities of individuals. Breeding birds have had to survive a period of severe shortage (usually the previous winter) or the rigours of migration. In less seasonal environments it is easier to survive to breed, but harder to find extra food for nestlings and fledglings. In the far north conditions may be severe even in summer and the start of the breeding season so late that clutch sizes are actually lower than farther south. For instance Orell and Ojanen (1983) found no correlation between clutch size and latitude in European Great Tits.

Marchant found no evidence of a latitudinal gradient in clutch size in his analysis of Yellow Robins, Willie Wagtails and Welcome Swallows from the RAOU Nest Record Scheme (Figure 26). Yom Tov (1987) on the other hand indicated that species in northern Australia tended to have smaller clutches than related species in the south. Unfortunately there are insufficient data from most species in the tropics to see if a true latitudinal gradient exists.

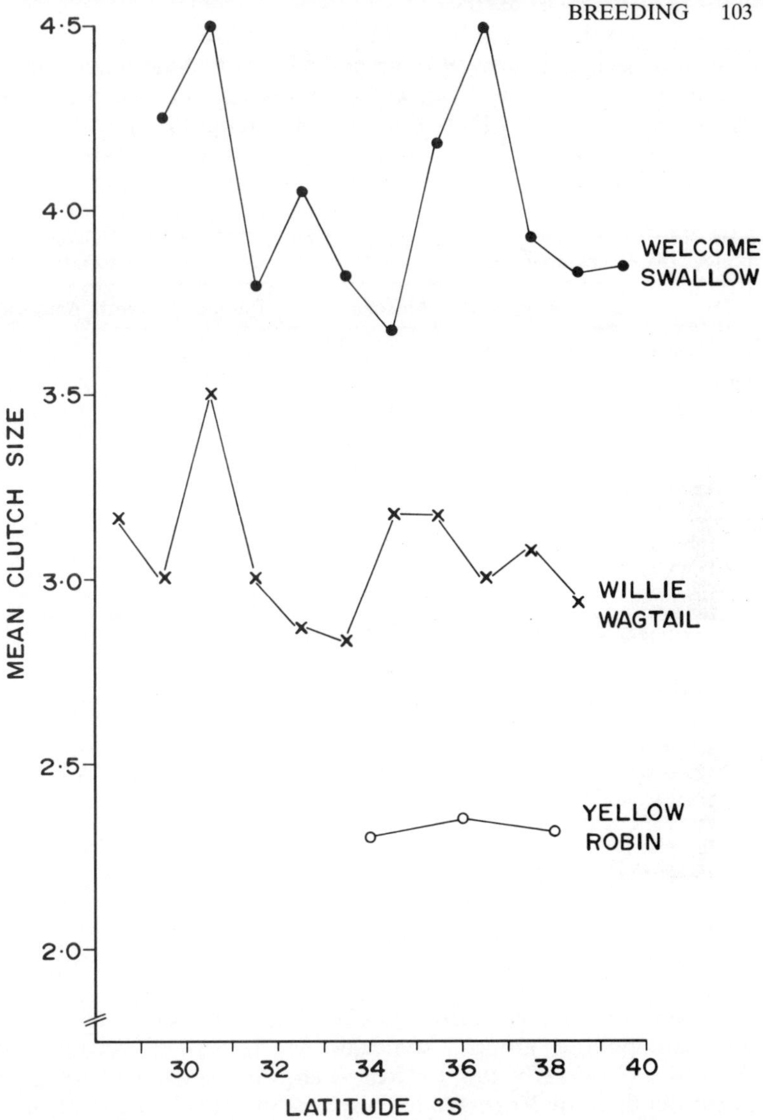

Fig. 26. Latitudinal differences in clutch size in three Australian passerines (from Marchant 1974, 1984 and Marchant and Fullagar 1983). None of the trends is significant.

The failure to find a latitudinal gradient in clutch size, and also the lack of substantial differences in the success of different clutch sizes could result partly from the narrow range of variation and low average size of clutches in most Australian birds. Woinarski (1985b) and Yom Tov (1987) compared clutch sizes of Australian birds with their relatives or ecological equivalents in the Northern Hemisphere. For most groups smaller clutches occur in Australia (Table 22). In the raptors, ducks, herons, crows and

Table 22. Clutch sizes in several groups of birds in Australia compared with ecologically similar groups in Europe and North America (from Woinarski 1985b).

	Australia	Europe	North America
Australian species related to Northern Hemisphere forms.			
Owls	2.8**	5.0	4.2
Crakes and Rails	5.8**	7.6	8.6
Raptors	2.8	3.6	2.9
Herons	3.6	4.1	4.3
Ducks	8.0	8.4	9.5
Pigeons	2.0	2.0	1.7
Corvids	4.5	5.4	4.4
Swallows	4.5	3.9	4.6
Finches	5.1+	4.9	4.1
Australian species unrelated to Northern Hemisphere forms.			
Frugivores	2.3*	5.5	3.2
Flycatchers	2.5**	4.8	3.4
Ground insectivores	2.8**	4.9	4.0
Foliage insectivores (large)	2.3**	6.1	4.5
Foliage insectivores (small)	3.1**	6.7	4.7
Large seed-eaters	3.6+	2.0	2.9

* Indicates Australian species have a lower clutch size than those of either Europe or North America.

** Indicates lower than both Europe and North America.

+ Indicates Australian species have larger clutch sizes than European or North American species.

swallows there are no differences between continents. These are all cosmopolitan groups, some of which arrived recently in Australia (Yom Tov 1987). Groups showing smaller clutches in Australia than the Northern Hemisphere mostly belong to ancient endemic families and are mostly insectivorous or nectarivorous. The seed-eaters (finches and a heterogeneous group of large species) actually have larger clutches in Australia. Woinarski (1985b) suggested that Australian environments are not markedly seasonal. They are mostly evergreen, have mild winters and insects display neither a great shortage nor a peak in abundance during most years (Woinarski and Cullen 1984). Conditions are simply not conducive to feeding many extra mouths even at the best times of year. Yom Tov (1987) presented a similar argument but proposed that the small clutches are a hangover from Australia's more equable past. It seems surprising that, if conditions have recently become more seasonal, birds were not able to increase clutch sizes. As noted earlier fairy-wrens alter the sizes of their clutches between good and poor years. In addition very recent arrivals, such as the Goldfinch (*Carduelis carduelis*), now have smaller clutches in some parts of Australia than their ancestors in Britain (Frith 1957; Middleton 1970b) and the same is true of the many exotic birds in New Zealand (Niethammer 1970). I believe that the question of small clutch sizes in Australian birds needs to be answered in the context of the present environment.

Ricklefs (1980b) used actual evapotranspiration (AE) as a measure of seasonal productivity. Highly seasonal environments have a high ratio of peak AE to lowest AE (usually in winter), which leads to high mortality and low density in the breeding season and consequently to an excess of resources. There was a negative correlation between minimum AE and clutch size when several very different sites were compared. Koenig (1986) reached the same conclusion from an analysis of North American woodpeckers. Cody (1966) proposed that low seasonality of resources leading to high breeding density and consequent low clutch size was an important contributor to the latitudinal gradient in clutch size. He also proposed that it explained other differences in clutch size. For instance birds of tropical savannas in Africa, which experience wet and dry seasons, have larger clutches than those in the more equable rainforest (Moreau 1944). In addition, birds on islands show low clutch sizes. In the European Robin, populations in the Canary Islands have by far the smallest clutches, and British birds have smaller clutches than those in Germany (Lack 1965). Islands are buffered from extreme temperatures and aridity by the surrounding ocean and their birds experience less fluctuation in resources (Blondel 1985). The low clutches in Australia and perhaps the Southern Hemisphere generally, could result from the large area of sea in the south and its modifying influence on the seasonality of the climate.

Is there any evidence of differences in clutch size between habitats in Australia, in particular larger clutches in the inland versus coastal areas? Welcome Swallows and Willie Wagtails have larger clutches in upland sites (presumably more seasonal) than coastal sites, whereas Welcome Swallows had slightly larger clutches inland than on the coast (Table 23). Comparisons between closely related species may also reveal slightly higher clutches in the inland species. The possibility of consistent differences in clutch size in different habitats within or between Australian bird species needs to be tested more extensively. It is

Table 23. Clutch sizes in different habitats at comparable latitudes in some Australian passerines.

	Coastal	Upland	Inland
Willie Wagtail (Marchant 1974)	3.00	3.22*	2.91
Welcome Swallow (Marchant and Fullagar 1983)	3.88	4.16*	4.18*
Yellow Robin (Marchant 1984)	2.31	2.33	+
Miners (Ford unpublished)	2.81[1]		3.17[2]*

* Significantly greater than clutch size on coast ($p<0.05$).
1 Noisy Miner (*Manorina melanocephala*).
2 Yellow-throated Miner (*M. flavigula*).
+ Uncommon inland.

possible that such differences account for, or are more important than, latitudinal differences.

Another unusual feature of some Australian passerines is the tendency to lay successive eggs in a clutch at intervals of about 48 hours (Marchant 1980). Most birds lay their eggs at 24 hour intervals. An extended laying period could result from a shortage of resources for egg-formation. Given the high predation rate on nests in Australia, there would seem little advantage in delaying successive eggs unless this is unavoidable.

Breeding Success

Even experienced birds that are breeding at the best season of the year and have laid the optimal clutch size will still have to overcome numerous difficulties to rear their offspring to independence. The most significant of these are predators and brood parasites, extreme weather and unexpected food shortages.

Predators include other birds (crows and currawongs are perhaps the most notorious in Australia), reptiles like snakes and goannas, and mammals from *Antechinus* to man. Australia has 12 species of cuckoos, all but one of which parasitize the nests of a range of landbirds. They remain virtually unstudied. Sudden cold spells in spring may chill eggs or kill chicks, especially as adults often have to spend long periods feeding themselves. Hot weather can also cause mortality and may be a factor in limiting summer breeding. Heavy rain and strong winds also play their part in swamping or dislodging nests and their contents or making it difficult for birds to incubate.

Nest success can be measured in several ways but it is most frequently presented either as the number of nests from which at least one young fledges or as the number of eggs that lead to fledglings. The first is easier to measure as one can include nests whose contents are not visible, but is less reliable as nests which do not receive eggs may or may not be included. Rearing young to independence, or recruitment into the breeding population, are the ultimate aims, but are difficult to measure especially in migrant species.

Breeding success of several Australian species is presented in Table 24, along with some representative data from Europe and North America. Success may exceed 50% in open-nesting species in Britain. Part of the reason for this may be the artificially low abundance of natural predators. In the Bialowieza Forest in Poland, one of the few remaining natural forests in Europe, a wide range of predators ensure that few attempts by birds are successful (Tomialojc *et al.* 1984). In Blackbirds (*Turdus merula*), semi-natural woodlands are apparently a more dangerous place in which to nest than are gardens (Snow 1958). Likewise in North America success in deciduous woodland can be very low (see Table 24). In Australia nesting success covers a similar range to that shown in Europe, though in some studies it was notably low (miners, thornbills). Perhaps data from the Nest Record Schemes, which pool results from many years and sites, are more reliable than those based on single sites in one or a few years. If so, there seems to be no reason to conclude yet that Australian birds are any less successful on average than their European or North American counterparts.

Predation is by far the major cause of losses in all three continents, with 80% or more of failures resulting from it in several studies.

Starvation is less frequent, and indeed quite unusual in Australian birds. Cuckoos can be a major factor, for instance Bronze-Cuckoos (*Chrysococcyx*) caused the failure of 20% of Splendid Fairy-wren (*Malurus splendens*) nests in Western Australia

Table 24. Breeding success of some Australian birds, with some values from Europe and North America, calculated from (i) percent of nests from which at least one young reared and (ii) percent of eggs leading to a fledgling.

	(i) % nests	(ii) % eggs
Australia		
+ Willie Wagtail (Marchant 1974)	64.9%	58%
+ Welcome Swallow (Marchant and Fullagar 1983)	65%	52.7%
+ Yellow Robin (Marchant 1984)	35%	28%
Thornbills (Bell and Ford 1986)	18–20%	
* Rosellas (Wyndham and Cannon 1985)	44–50%	41%
Fantails (Cameron 1985)	59–71%	
Fairy-Wrens (Rowley 1965, 1981)		54%
Noisy Miner (Dow 1978a)	15%	15%
Chats (Williams 1979)		37%
Europe		
+ Thrushes (Snow 1955)	40–55%	
+ Warblers (Mason 1976)	55–68%	56–65%
* Titmice (Perrins 1979)	70–80%	60–70%
+ Chaffinch (Newton 1964)	44%	41%
+ Chats (Fuller and Glue 1977)	35–60%	19–38%
+ Nightingale (Morgan 1982)	69%	
North America		
11 species (Nolan 1963)	21%	18%
Many species (Nice 1957)	49%	

* hole-nesting species.
+ based on Nest Record Schemes.

(Rowley 1981). The weather can be an important direct factor. Honeyeaters desert many of their nests during periods of heavy rain or when there are frequent strong winds and thunderstorms (Dunkerley, pers. comm.; Marchant, pers. comm.).

Summary

As indicated in the introduction the many facets of the breeding cycle are closely related. For instance, birds need to make compromises between how often to breed and how many eggs to lay in a clutch. John Woinarski (1985b) has drawn together the different aspects of breeding biology of Australian birds and contrasted the pattern with that of the temperate Northern Hemisphere. His emphasis has been on insectivorous birds.

He found that, compared with their northern counterparts, Australian species have long breeding seasons in which they make repeated nesting attempts with small clutches. The nesting cycle may be slightly longer as incubation and nestling periods are a day or two longer than comparable species in Europe and North America. This adds little to the length of the breeding season though. Breeding success is perhaps no lower than on other continents, but with the small clutches, the productivity per breeding season is low. Small birds also appear to be longer-lived in Australia (data on adult survival and longevity are presented in Chapter 8).

Woinarski attributed the more leisurely and prolonged breeding of Australian insectivorous birds to the low seasonality of their food. Several studies now have indicated that whereas insects on foliage and elsewhere tend to peak in spring or summer they are still reasonably plentiful in winter, especially in drier habitats (Woinarski and Cullen 1984; see also Chapter 4). This is probably because eucalypts, the predominant trees in forests and woodlands, are evergreen and indeed produce fresh leaves almost throughout the year. Deciduous woodland in Europe and North America has a brief period of foliage growth and associated burst of insects, and the evergreen coniferous forests are also highly seasonal, often with harsh winters. The evergreen nature of southern Australian habitats and long growing season result from the rather equable climate, itself a consequence of the large body of water surrounding the country and indeed in the Southern Hemisphere as a whole. The Northern Hemisphere has large continental blocks, which, except on their fringes, experience inhospitable winters that impose high mortality on the few species that do not migrate. Migration also exacts a toll on the mobile species.

As a whole, Australian insectivorous birds, and perhaps to a somewhat lesser extent other birds, experience an annual pattern of resources and display a response similar to that shown in Figure 22b, rather than the more typical temperate pattern of Figure 22a. Another way of expressing this is to say that they are more "K"-selected than their "r"-selected northern counterparts.

There is also apparently a geographical uniformity in the breeding behaviour of Australian birds. A spring breeding peak is displayed in the cool temperate south and the monsoonal north, albeit for different reasons. Latitudinal gradients in clutch size within a species are yet to be demonstrated. This may be because the breeding peak is close to the vernal equinox or because annual patterns of resources are similar throughout the continent.

This equable environment and rather restrained breeding is in marked contrast to the picture that is often portrayed of Australia as a land of extremes to which birds respond in an opportunistic pattern. This change in emphasis results from the detailed accounts of the breeding biology of several species from coastal areas. We still require detailed studies of the breeding biology of birds from central Australia.

6. Mating Strategies and Parental Behaviour

The previous chapter dealt with the ways in which breeding birds respond to the environment and the resources that it provides. In this chapter and the next one I shall describe and compare individual species to show the diverse ways in which they breed. I shall attempt to provide the ecological reasons for these differences. I shall concentrate on the behaviour of the parents, in particular how they respond to each other as mates and to their offspring. The mating strategy adopted by a breeding bird is intimately related to its subsequent contribution to the rearing of its offspring.

Mating Systems

Monogamy is by far the most common mating system in birds in all parts of the world. It involves a single male and female, between which a pair-bond is developed. The pair remains together during the breeding attempt, usually the whole season and in some species for life. Both parents contribute to rearing the young though there may be some division of labour. The male often plays the major part in territorial defence, whereas the female builds the nest. He may help to feed the female when she is laying eggs or incubating. The female alone or sometimes both incubate the eggs and both bring food to the nestlings and fledglings and guard them. Monogamy is the rule among seabirds, raptors, parrots and pigeons and is common in the passerines. Parental care is far better developed in birds than in most fish, reptiles and amphibians and generally when two individuals contribute, success is greater than when one alone has to rear the offspring. There is usually little incentive for either sex to desert the other and seek additional mates. This is not to say that individuals do not attempt to copulate with birds other than their mates. Indeed detailed studies have shown that this occurs quite frequently in many species (Birkhead *et al.* 1987). Males of many species guard their mates to prevent becoming cuckolded. In mammals the female has, through pregnancy and lactation, taken over the major parental role in most orders, so that monogamy is much less common than amongst birds.

There are numerous examples in birds though, where one sex, typically the male, has reduced its contribution to the care of offspring, but on the other hand enhances its fitness by mating with additional partners. The evolution of *polygamy*, as such multiple mating is called, depends upon the extent and type of parental care given by a species. Where some pair-bond is developed it is termed *polygyny* (one male and several females) or *polyandry* (one female and several males). Where the sexes only come together for mating it is called *promiscuity*. The pattern of dispersion and abundance of food and other resources such as nesting sites also influences whether monogamy or polygamy develops. The ecological and behavioural basis for the evolution of mating systems, in particular how polygamy evolved from monogamy, has been extensively discussed in recent decades. Halliday (1978, 1980a) and Emlen and Oring (1977) provide excellent reviews of the subject.

There are numerous circumstances under which polygamy may be a more favourable option than monogamy, at least for the male. These fall into four major groupings on ecological grounds (Table 25). Polygamous species have previously been separated into those in which male success is determined by female choice and those in which males actively compete for territories or display sites where they can gain access to or defend the females. This division has been found to be inadequate, as most examples involve both. Even in typically monogamous species a few individuals may manage to obtain a second mate. Parental care is divided between the ensuing broods, often unequally with the first female gaining most help. Pied Flycatcher (*Ficedula hypoleuca*) males arrive in spring in northern Europe and immediately start to defend nest holes (Alatalo *et al*. 1985). Females arrive a few days later and choose a male and a nest hole. Some males that obtain mates early may then start to defend a second nest site some distance away and may attract a second female. The second female suffers poor breeding success that season, because the male provides more assistance to his first mate. Once the second female has begun to nest it is probably too late to find another mate and nest hole. Probably the situation among Pied Flycatchers is now atypical as so many use nest-boxes. Natural holes in trees were

Table 25. A classification of the main types of polygamy in birds.

Reason:	Habitat	Food	Male competition	Female choice	Examples
Opportunistic	Woodland etc.	Insects etc.	+	?	Pied Flycatcher
Variable territory quality	Marshes, Grassland	Insects etc.	++	+	Red-winged Blackbird Cisticola Rufous Songlark?
Precocial offspring	Mostly open	Adult: plant material or invertebrates Young: emergent insects	++	++	Grouse Some waders
Emancipated Male	Tropical forest	Fruit, nectar, plus insects in young	++	++	Manakins, Cotingas, Bowerbirds Birds of Paradise Hummingbirds

originally more scarce, so it would have been difficult for a male to defend more than one in a season.

Such opportunistic polygamy in species where most individuals are monogamous is apparently not uncommon, especially among long distance migrants where time to pair up and breed is short (Møller 1986). Further detailed studies on colour-banded birds, may reveal how frequently it occurs.

In a few passerines polygamy has become the rule. Many of these occupy rather simple environments, where resources are abundant but patchily distributed, so that exceptionally good territories can be defended. In Europe and North America polygamy is found among several marsh and grassland species, including some warblers (Sylviidae, Cisticolidae), icterid blackbirds and wrens (Troglodytidae) (Verner and Willson 1966; Orians 1980; Møller 1986). Males defend territories aggressively and vocally, and some obtain high quality areas where food is abundant and nesting opportunities good. Females can rear young by themselves, although they are more successful if they receive assistance from their mate. Theoretically polygamy can be explained if one considers that the females choose their nesting sites sequentially. The first females should choose the best sites, where food, such as emerging insects, are abundant and there is dense cover. The next females have a choice of being second in a good territory or being the first or only female in a poorer territory. At some point, as vacancies fill, the former option will be better. Exactly when this occurs depends on the difference in quality between the territories and how important paternal care is (Figure 27). The breeding biology of marsh-dwelling and grassland birds in Australia has been virtually neglected. To date the Rufous Songlark (*Cincloramphus mathewsi*) has been identified as polygamous (Ryan 1954). It is interesting that it belongs to the old world warblers (Sylviidae) where polygamy is known to be frequent. Other warblers in Australia, Brown Songlark (*Cincloramphus cruralis*), Grassbirds (*Megalurus*) and Cisticolas (*Cisticola*) are likely candidates for polygamy (Chisholm 1953).

A common feature of polygamous species, including the Rufous Songlark, is that males are much larger than females. Occasionally, as in Red-winged Blackbirds (*Agelaius phoeniceus*), males are also more brightly coloured (Orians 1980). Competition among males is intense and selection has favoured an increase in size as larger birds are more likely to gain high quality territories. As such selection is related to breeding success rather than survival

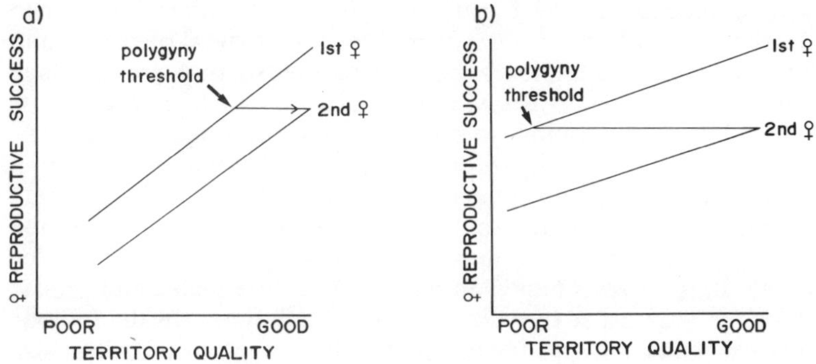

Fig. 27. Relationship between the likelihood of polygamy, female reproductive success and territory quality. Females are expected to choose the best quality territory that is available. Higher quality may mean safer nest sites or more food or both. As good territories are occupied a point may be reached (the polygamy threshold) where a female will be more successful as a second mate in a good territory than as the only mate in a poor territory. This is more likely in case (a) where there is a wide range of quality of territories or abundant food so that male help is not essential. This would be true in Red-winged Blackbirds. In case (b) there is less disparity in territory quality and help from the male appreciably improves the female's success. Most woodland birds would be like this. (Model modified from the work of Orians 1969, 1980).

it was termed sexual selection by Charles Darwin (1871). Whereas it is sensible to consider breeding and survival separately and indeed strong sexual selection may lead to reduced adult survival, sexual selection is only really a component of natural selection. As will be seen in the next two groups sexual selection can be extremely powerful and can produce bizarre results.

Not all young birds are helpless at hatching (*altricial*) and remain in the nest (*nidicolous*). In the galliforms and most wading birds the chicks are well developed (*precocial*) and leave the nest within a day or two of hatching (*nidifugous*). The parent or parents guard their young from predators and generally take them to areas where food is plentiful. The chicks often eat different food from the parents, typically emerging insects which can be very abundant. Emancipation from parental duties has been common in such nidifugous birds. Males, and occasionally females, can spend much of their time displaying to potential mates. These may occur on solitary display grounds (*arenas*) or even communal ones (*leks*). Amongst many species of grouse the males congregate at leks and display to the females who wait around the periphery. Females exercise a degree of choice in selecting a mate. There is strong sexual selection for elaborate plumage and displays. European Black Grouse (*Lyrurus tetrix*) undergo a noisy ritual of raising and lowering wing and tail feathers, whereas the North American Sage Grouse (*Centrocercus urophasianus*) releases air from a sac in its throat to make popping noises. Although the females make the ultimate choice they apparently are guided by a male's position in the lek, as much or more than by the display itself. Males compete for and, with age and experience, move gradually towards the centre of the lek where most of the mating takes place (Wiley 1978). Large size is hence an advantage to the males and grouse show the greatest sexual dimorphism of any birds. In the lekking species, males play no further part in breeding, indeed their size and conspicuous plumage make them a liability anywhere near the nest or chicks.

The wading birds show a remarkable diversity in the breeding system. In some species the male and the female both help to guard and guide the chicks (Walters 1984). In others such as the Sanderling (*Calidris alba*) the female lays a clutch of eggs which the male looks after and a second clutch which she herself incubates (Parmelee and Payne 1973). This is taken a stage further in the American Jacana (*Jacana spinosa*), where the female mates with a succession of males and leaves each with a clutch of eggs to incubate (Jenni and Collier 1972). Female Jacanas are larger than males and defend large territories. There are similarities between their behaviour and that of male Red-winged Blackbirds. The Red-necked or Northern Phalarope (*Phalaropus lobatus*) is one of the few other polyandrous birds. In this species females are more brightly coloured than the males and gather at communal display and mating grounds. There is some bond between the female and each of her successive males, probably just long enough for her to lay a clutch of eggs. Males assume the parental responsibilities.

Perhaps the most remarkable of the waders is the Ruff (*Philomachus pugnax*). Males are large and adorned with ruffs and head tufts of white, brown and black. They assemble at leks to which females are attracted. To complicate matters Ruffs are polymorphic in both colour and behaviour. Those with dark feathers try to defend a small territory in which they display and generally attract most of the females. Other white-ruffed birds are less aggressive but are tolerated on territories by the darker birds. These satellite males may help to attract females to a territory and so enhance the fitness of the holder, as well as their own fitness as they sometimes mate successfully. If only what is happening at a single lek is considered the dark, territorial birds appear to do best. However, white-ruffed birds may move between leks and possibly even migrate with the northward-moving females. White males are more common in Scandinavia than in the south of the species' breeding range. Probably on average the two colour forms are equally successful so that they are retained at a reasonably constant frequency. Were the dark territorial forms to predominate perhaps the few white individuals would become more successful. If the satellites became too common then there would not be enough territories to join. The Ruff's complicated mating system has been unravelled by long-term studies in the Netherlands (Hogan-Warburg 1966; van Rhijn 1983), but there are still parts of the story that remain speculative.

Another main group of polygamous species is that in which the adults rely on an abundant high carbohydrate, but low protein, diet. The manakins and cotingas of tropical America feed on fruit and the males can feed themselves in a short proportion of the day. This leaves plenty of time for display (Lill 1974; Snow 1976). Young birds require far more protein than adults and are fed on insects which are less abundant but more evenly distributed than fruit. The female is able to find enough insects near the nest to rear the young herself, and can feed herself rapidly on fruit. Although the female could perhaps rear more young with help from the male, his best strategy is apparently to concentrate on displaying rather than spending time on parental care.

The prevalence of polygamy in frugivorous species is paralleled in Australasia by the bowerbirds and birds of paradise. These are among the most spectacular arena birds and research on them is contributing greatly to an understanding of the evolution of polygamy and its ecological correlates. They will be discussed in detail later in this chapter. Many hummingbirds, which rely on nectar that has even less protein than fruit, are polygamous by defending rich territories (Wolf and Stiles 1970). It is interesting that polygamy has not yet been recorded in the honeyeaters, which rely on nectar, fruit and other foods high in carbohydrates. Some species, such as the sexually dimorphic Crescent Honeyeater (*Phylidonyris pyrrhoptera*) and the Tawny-crowned Honeyeater (*P. melanops*), which displays in flight over open heathland, would appear to be potential candidates for polygyny.

Monogamous Species

Rufous Whistler

With so many exciting polygamous species and co-operative breeders, straight-forward monogamous birds have been rather ignored in Australia. One notable exception is the Australian Raven (*Corvus coronoides*), studied by Ian Rowley (Rowley 1973a,b). Another, which I have chosen to discuss in detail, is the Rufous Whistler (*Pachycephala rufiventris*) (Plate 34). In many ways it is a bird whose breeding behaviour would be unremarkable in North America or Europe.

Rufous Whistlers have been studied near Armidale for several years, principally by Lynda Bridges. They are summer visitors, with the males arriving in late September and October. Males are more brightly coloured than females and sing frequently and vigorously throughout spring. The song appears to function both as a deterrent to intruders into the territory and to attract females. Males often counter-sing when in neighbouring territories and may display against each other and even fight. Wandering males are driven out of territories. Females arrive shortly after the males and soon pair up with a male. They may be paired to the same mate in successive years. It should be noted that sexual selection can still operate in monogamous species. It appears that some male Rufous Whistlers are unsuccessful in obtaining a mate, but it is not known what characteristics influence the females' choice. Male Rufous Whistlers take at least two years to attain adult plumage and once they have a territory they may return to it for several years (Bell and Ford 1987). In addition, and curiously, a number of males in the dull, female-like plumage sing, defend territories and may even breed successfully. It is unclear whether these are first year birds or older birds that fail to attain the full plumage. Such breeding by immature birds is shown in several other dimorphic passerines, including Pied Flycatchers. In the latter case some brown males never attain the black and white plumage characteristic of most males.

Only the female builds the nest, though the male may help to select the site. Incubation and feeding of the nestlings and fledglings is almost exactly evenly shared. Both clutch size (mean 2.4) and breeding success (0.45 young/nest) are low, with high predation rates, but they make repeated attempts. Rufous Whistlers take rather large insects (beetles, larvae, cicadas) for their size (about 25 grams) and feed their nestlings (6.5 visits to nest/hour) and fledglings (5.7 visits per bird/hour) rather infrequently. One can imagine that a single parent attempting to rear young would experience considerable difficulty.

Polygamous Species

Lyrebirds

In stark contrast to the female Rufous Whistler, the female Superb Lyrebird (*Menura novaehollandiae*) rears her young with no assistance from her mate. Lyrebirds do not appear to fall into any of the normal categories of polygamous species. Food is probably not particularly abundant, they feed on invertebrates unearthed from leaf litter and the young are by no means precocial. However, the male has become freed from parental duties and spends much of his time in the breeding season attempting to attract females.

Lyrebirds occur in dense forest along the southeastern coast and ranges of Australia. They have been studied in a casual way in many areas but those of Sherbrooke Forest near Melbourne are the best known (Chisholm 1960; Reilly 1970). For such large birds they are surprisingly difficult to observe clearly, but birds in Sherbrooke Forest have become used to people to some extent.

Lyrebirds nest in winter and for several months the males perform an energetic and spectacular audio-visual display from mounds within their territories. The song is so powerful that some phrases can be heard from several kilometres away. It is only at close quarters though that its full complexity can be appreciated. The male Lyrebird intersperses its own rich and melodious notes with the calls of other species, such as black-cockatoos, rosellas, shrike-thrushes and currawongs. Individual birds may mimic a dozen or more species in the course of a few minutes. Robinson (1974) has suggested that most of the birds that are copied do not breed in the winter so that they will not be singing. However, many of the calls are not necessarily associated with breeding, e.g., those of rosellas and cockatoos. Robinson (1974) also noted that many of the models are rather large, often aggressive or even predatory birds. It is true that goshawks, kookaburras and currawongs are copied, but so too are scrubwrens, whipbirds and some honeyeaters. Perhaps the mimicry of other birds is simply a means of extending the repertoire of the songster. An even simpler hypothesis is that mimicry simply represents an accumulation of mistakes (Hindmarsh 1986).

Surprisingly, the proof that Lyrebirds are polygamous was only provided recently (Kenyon 1972), and there is, as yet, little information on the relative success of different birds, or indeed whether females choose between males or simply mate with the male whose territory they occupy.

A broad repertoire in other mimetic birds, such as Northern Mockingbirds (*Mimus polyglottus*) (Howard 1974), and Reed Warblers (*Acrocephalus scirpaceus*) (Catchpole 1980) has been linked with success in attracting a mate, though these are monogamous species. In the Great Reed Warbler (*A. arundinaceus*), males with one mate have a wider repertoire than those with no mate, those with more than one mate have a still wider repertoire (Catchpole 1986). Birds with a wide repertoire also leave more offspring in each breeding season. Possibly the repertoire of individual Lyrebirds is related to the number of matings he will achieve. Conceivably, older birds, which have proved that they possess good genes by surviving, have learnt to copy more species of birds. Such speculation can only be tested by observation at numerous display mounds over many years.

The male Lyrebird's display does not end with interspecific mimicry. When a female is close to a singing bird he will start his visual display, in which the lyre-shaped tail is arched over the back and vibrated vigorously (see Plate 17). The song changes markedly during this display with less mimicry and more calls that are specifically the Lyrebird's. A "cric-cric" call is frequent and birds may make a continuous mechanical whirring noise.

Lyrebirds give every impression of sexual selection being stretched to its limits. Apparently the females require elements of the typical Lyrebird calls to be attracted near enough to hear the full song. She needs to sample the full repertoire before coming close enough to experience the visual display, along with its sound effects. Only then may she decide to accept the male as a mate. Perhaps there is good reason for her demand for quality. She receives no help from the male (Lill 1979); indeed all that he gives her offspring are his genes and it is to her advantage to choose the best that are available.

Just how hard the female has to work to rear her single offspring has been unravelled by the painstaking work of Alan Lill (Lill 1986). The nest weighs about 14 kg and takes about 140 hours to construct. The single egg is incubated for about 50 days during the coldest period of the year, and the nestling remains in the nest for nearly as long again. Nestlings are fed about 16 times per day, mostly on earthworms and amphipods. The fledglings seem to be dependent on their mothers for a further 8-9 months, in other words until the start of the next breeding season. They are fed much more frequently than nestlings and Lill (1986) estimated that a fledgling may receive approximately 26,669 meals before reaching independence!

The very long incubation period is because the female has to leave for extended periods during the day in order to feed herself. The long nestling period results from very slow growth due to the rather few feeds per day.

If rearing a young Lyrebird is such hard work and takes such a long time, why does the male not help? Lill believes that three factors have probably favoured lack of male parental care. Male Lyrebirds occupy large territories which are densely vegetated and they would find it very difficult to stay with the female while she is fertile and at the same time attend to territorial defence. If he did not there would be considerable opportunity for other males to sire the single chick; in such circumstances it is difficult to see how there could be selection for paternal care. Second, allowing for the fact that the male's contribution would probably be less than the female's because, for instance, he is responsible for territorial defence, both birds may still be unable to rear more than one young. In other words, if the male helps in breeding, the potential number of young still may not be increased. Finally, egg-laying is probably sufficiently asynchronous to give a male, once he has mated with one female, a reasonable chance of mating with a second female. This would immediately double the potential number of offspring that he could father. Male Lyrebirds attempt to be polygamous and therefore desert the females; by doing so they increase their own fitness, regardless of what their behaviour does to the fitness of the female or indeed the species as a whole.

Bowerbirds

Bowerbirds are undoubtedly amongst the most remarkable of birds, and have attracted attention from the time of John Gould onwards. There are 18 species, restricted to Australia and New Guinea, of which 14 display in arenas. Some of their displays were described by Marshall (1954), but it is only in the last decade that one species in particular, the Satin Bowerbird (*Ptilinorhynchus violaceus*), has become well understood (Vellenga 1970, 1980a,b; Diamond 1982; Borgia 1985a,b, 1986).

Reta Vellenga and her husband, working from their garden in the Blue Mountains unravelled the complexities of the Satin Bowerbird's way of life. They provided an excellent background from which Jerry Borgia could propose and test specific hypotheses experimentally.

Adult males are glossy blue-black with a purple iris and yellow beak, but it takes them about six years to develop this plumage (Plate 35). Young males and females are green and brown. The males start to build rudimentary bowers well before they attain the black plumage, but require several years of practice to perfect the technique. Bowers are constructed in small clearings, usually where the morning sun reaches the ground. Sticks are laid on the ground and in a double row stuck vertically into the ground so that the avenue between them faces approximately north-south. The occupant adds ornaments around, but not in, the bower. These are yellow, brown and in particular blue and include snail shells, yellow leaves, flowers and feathers. With the untidy ways of Europeans the availability of decorations has increased with drinking straws, bottle tops, coloured paper and clothes pegs being most frequently collected. As if this is not enough some bowerbirds even paint their bowers with chewed up fruit or charcoal and saliva. Bowers are in traditional sites that may be used for several decades, perhaps by the same individuals. Male Bowerbirds need to spend a great deal of their time at their bowers not only to construct them and display to potential mates, but also to prevent other males from wrecking the bower and stealing the ornaments. The Vellengas showed that bowers could be destroyed very rapidly if left unattended. They also found that there was a dominance hierarchy with dominance associated as much with a site as with an individual. When their dominant bird disappeared his bower was taken over by the next ranked bird.

Females visit several bowers and, if they can withstand the aggression of the holder, may mate with him. Mating attempts can be disrupted by other males. Although there is a clear element of female choice in the relative mating success of the males, the males clearly compete strongly amongst themselves for the best sites and decorations. Bowerbirds are frugivorous as adults, though the females, who have sole responsibility for rearing the young, feed their offspring on insects.

Borgia worked on a Satin Bowerbird population near Tenterfield in northern New South Wales. He was able to ask specific questions on the function of the bower by a combination of automatic photography and experimentation. Clearly sexual selection has been a major force in the development of the bower, but why has it become so complex and what is the significance of the ornaments? There are several possible explanations of how selection has favoured the evolution of the bower. At the simplest level females may move about randomly and happen to mate with males that display for the longest time each day or have bowers in the most conspicuous sites. Alternatively females may select the males with which to mate. The favoured males should have the best genes for two rather distinct reasons. First, females may select males that possess characters that could be passed on to both sexes of offspring. These include good ability to escape predators and high foraging efficiency. Older birds, simply from the fact that they have survived so long, provide evidence of their genetic quality. Second, females may choose the most attractive males because his attractive features will be passed on to her sons, who should likewise prove attractive to the females of their generation. Cruder aspects like fighting and stealing will be favoured, along with characteristics like the skills of bower-building and finding rare ornaments. This sounds like a circular argument, and indeed it does involve positive feedback. Females choose the best males, because the best males attract the most females! This is the "run-away hypothesis" of Fisher (1930).

A modification of the hypothesis that females select the males which probably have the best genes, is the handicap hypothesis of Zahavi (1975). The very fact that the male is brightly coloured and spends much of his time displaying at a conspicuous bower puts him at risk from predation and may cause him to waste time that he could well spend in feeding. It's as if he is saying that he must be able to look after himself well if he can spend so much time hedonistically. The problem with this hypothesis is that the poorer characteristics as well as the good could be passed on to the sons and the daughters would perhaps gain little. Finally, the female could simply be choosing healthy males that have bowers in safe places.

Unfortunately these hypotheses are unlikely to be mutually exclusive but Borgia's work gives strong support to the idea that females select mates that are likely to be genetically superior. For a start there is a great inequality in the success of males (Figure 28). The favoured male mated 33 times in one season whereas a few did not mate at all. Second, the most successful males had the neatest, most symmetrical bowers with the best collection of decorations. Blue rosella feathers in particular were a strong indication of a successful male. Borgia and his colleagues could manipulate the male's success by taking away feathers from the

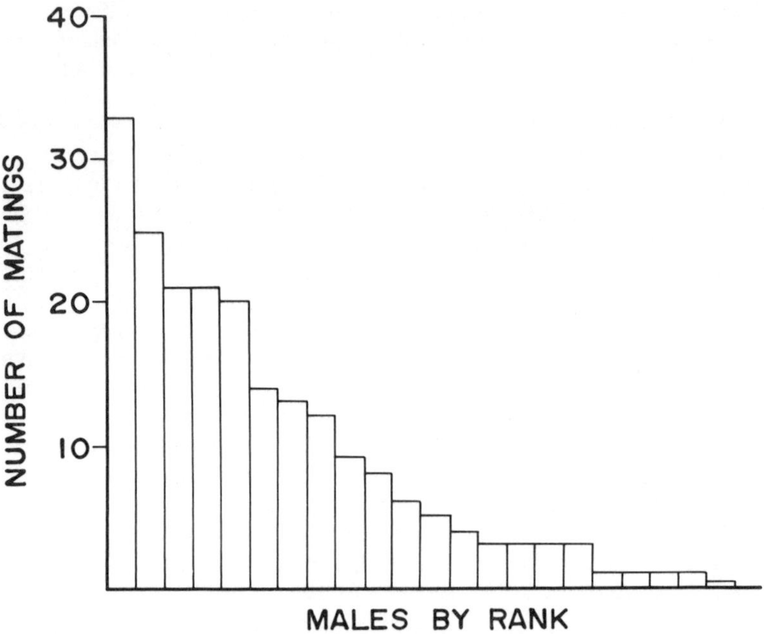

Fig. 28. Relative mating success of Satin Bowerbirds (modified from Borgia 1985a).

dominant birds and giving them to lower ranked birds. Females did not often mate with males that frequently had their bowers wrecked. Presumably males that can hold together a neat, well-decorated bower are advertising their quality, which gives an indication of their genetic worth. Probably both good genes for basic survival, plus those for the "run-away" characteristics are important.

Macgregor's Bowerbird *(Ambyornis macgregoriae)* has now received close study in the central cordillera of New Guinea (Pruett-Jones and Pruett-Jones 1982, 1985). The dull-brown males have an orange crest, which they erect while displaying. They erect maypole bowers regularly spaced along ridge-tops. These are decorated with charcoal, fungi, leaves, seeds, fruits and insect frass. Males display in the bower, partly hidden from the female, who is enticed in. If the male then rushes out the female will often leave. Occasionally she stays in the bower long enough for copulation to result.

If Borgia's and the Pruett-Jones' work goes some way to explaining why bowerbirds build bowers and display, what about the question of how these complex behaviours evolved? Diamond (1982) made some suggestions. The bower itself may be an extension of nest-building (though Borgia 1985a and Borgia *et al.* 1985 dispute this). In some species of birds, males build many potential nests to court the female and stimulate her into breeding condition. Also males often present the females with food during courtship. This could explain ornaments such as fruit and flowers that occasionally now may be extended towards a female at the bower. Some bowerbirds do cache edible fruit near the bower (Pruett-Jones and Pruett-Jones 1985). Most decorations are inedible, but possibly those inert objects could replace the edible items, if they stimulated the female. Painting of the bower is also perhaps an extension of courtship feeding, possibly a male involved in presenting food to the female became involved in nest-building as a displacement activity.

One of the most significant points about bowers and bowerbirds is that the dullest species appear to have the most elaborate bowers. The brilliant gold and black Regent Bowerbird (*Sericulus chrysocephalus*) builds a small simple bower whereas the Vogelkop Bowerbird (*Amblyornis inornata*) is plain brown but builds a "hut" one metre high. Gilliard (1969) proposed that the secondary sexual characteristics of male bowerbirds have been progressively transferred from brilliant plumage to the bower. This has advantages in that the dull male can slink away from predators, whereas his bower means that his impact can be magnified in the eyes of the female. It is interesting that the male Satin Bowerbird's favoured colours are blue, like his plumage and yellow, like his beak. He can enhance his image by surrounding himself by blue and yellow objects. An alternative possibility is that the male, during courtship, can redirect his aggression towards his coloured ornaments, as if they are surrogate rivals. This may make him less likely to drive away the female.

Borgia *et al.* (1985) suggested that the original bowerbirds cleared leaves and sticks from the ground to provide a display ground. Manipulation of the sticks and other objects led ultimately to the construction of bowers. The bowers, especially where they are hut-like could provide the male and female with some protection from predators during courtship and mating. Most important they are an indicator to the female of the quality of the male.

Although we are close to understanding the behaviour of the Satin Bowerbird, there are many questions to ask about the other species and one can expect more surprises. What determines the female's choice in these species? Do males interact aggressively with each other? Are species with simple bowers representative of a primitive stage or are they losing the habit? How rapidly do changes in behaviour evolve? How will changes brought about by humans to the environment affect the birds' behaviour and success? Are bowerbirds intelligent and do they display an aesthetic sense?

In relation to the last two questions certainly bower-building is a skill that has to be learned, and opportunities for improvement are rapidly exploited. If an aesthetic sense means an appreciation of beauty, then the female Satin Bowerbird appears to appreciate a neatly built and well decorated bower. Perhaps this is only because it tells her something about its owner; maybe as scientists we should regard the bower simply as a functional symbol of the quality and status of its owner rather than an article of beauty in its own right. But what then is a millionaire's private art collection?

Birds of Paradise

The birds of paradise rival bowerbirds as extreme examples of sexual selection. They do not construct bowers although some species do clear arenas on the forest floor. Lawes' Parotia (*Parotia lawesii*), which forms leks, also places ornaments like snake-skin, mammal scats and fur, chalk and bone around the arena. These may be taken by the female for the nest or as a mineral supplement (Pruett-Jones and Pruett-Jones 1988a). In contrast, birds of paradise excel in their spectacular plumage and visual displays. The 42 species are found in New Guinea and nearby islands and extreme eastern Australia. They are apparently not closely related to the bowerbirds but are recent offshoots of the crows. Their beautiful plumage, and in most species the contrast between the sexes can be seen in the recent book "Birds of New Guinea" (Beehler *et al*. 1986), whereas some examples of their displays are shown by Beehler (1983a) and Pruett-Jones and Pruett-Jones (1986). Birds of paradise are mostly frugivorous and promiscuous, males mating with several females with no pair-bond or paternal care. Thus, the pattern shown by the bowerbirds and the Neotropical frugivores (manakins and cotingas) is repeated. In fact, the birds of paradise show quite a variety of diets and mating behaviour and comparisons within the family have allowed Bruce Beehler (1983b, also Beehler and Pruett-Jones 1983) to examine more closely the link between frugivory and polygamy.

Snow (1976) argued convincingly that fruit in tropical forests is often abundant and indeed conspicuous. It can therefore allow males, in particular, plenty of free time to devote to display. In addition, most birds in tropical forests have small clutches, not necessarily because food is scarce but because predators are common and total losses frequent. The chances of success are not great so that modest reproductive effort is favoured and the female alone is able to feed the young.

Although this line of reasoning has considerable merit it is by no means the whole story. There are many tropical forest frugivores that are not polygamous, including some birds of paradise. Beehler (1983b) looked more closely at the diet and behaviour of nine species at Mount Missim in Papua New Guinea (Table 26). The genus *Paradisaea* probably includes the most

Table 26. Nine species of birds of paradise studied by Beehler (1983b).

Name	Breeding system	Food
Crinkle-collared Manucode *Manucodia chalybatus*	Monogamous	Figs
Trumpet Manucode *M. keraudrenii*	Monogamous	Figs
Magnificent Riflebird *Ptiloris magnificus*	Polygamous (solitary display)	Arthropods>Fruit
Black-billed Sicklebill *Epimachus albertsii*	Polygamous (solitary display)	Arthropods>Fruit
Superb Bird of Paradise *Lophorina superba*	Polygamous (tree display)	Fruit>Arthropods
Lawes' Six-wired Bird of Paradise *Parotia lawesii*	Polygamous (terrestrial court)	Fruit>Arthropods
Magnificent Bird of Paradise *Diphyllodes magnificus*	Polygamous (terrestrial court)	Fruit>Arthropods
Raggiana Bird of Paradise *Paradisaea raggiana*	Polygamous (lek tree)	Fruit> Arthropods
Blue Bird of Paradise *P. rudolphi*	Polygamous (solitary arena)	Fruit (incl. Figs)> Anthropods

spectacular members of the family, such as Count Raggi's Bird of Paradise (*P. raggiana*). Most are large with, in the males, immensely long and fluffy yellow to red plumes that extend from the flanks. These they throw over their heads in an almost inverted display to any approaching female. Typically this genus displays in small leks of two to about six birds. An exception is the Blue Bird of Paradise (*P. rudolphi*) which mounts a solitary display in which the iridescent blue breast feathers are fanned out as the bird clings beneath a branch (Pruett-Jones and Pruett-Jones 1988b). The parotias (*Parotia*) are somewhat less spectacular and display on a cleared dance court on the forest floor. Both of these genera are highly frugivorous, at least as adults, taking rich abundant fruits of a great variety. The glossy black, and sexually monomorphic, manucodes (*Manucodia*) are also frugivorous but are strictly monogamous. Beehler (1983b) argued that this is because they are fig specialists and figs are poorer quality food than most fruits. MacGregor's Bird of Paradise (*Macgregoria pulchra*) is also monomorphic, monogamous and a specialized frugivore, eating podocarp fruits (Beehler 1983c). Curiously the sicklebills (*Epimachus*) and riflebirds (*Ptiloris*) are markedly insectivorous, using their long curved bills to poke into the bark of forest trees (Bell 1984), yet they are still polygamous, with the males performing inverted displays (Beehler 1987). Have they only recently returned to a more animal diet or does fruit still provide a significant proportion of their energy?

Beehler (1983b) suggested that both fruit-eating and low clutches are prerequisites for polygamy in tropical forests but they do not necessarily lead to it. Fruit needs to be of good quality to allow males free time. Also, only species that have a generalized and rich diet are likely to occupy large home-ranges that enable individual males to encounter numerous females. Almost all species of birds in tropical rainforests have small clutches, birds of paradise typically have two eggs, and there is no obvious difference between monogamous and polygamous species. There is a need to look in more detail not only at the diet of frugivorous birds, but also at its quality and how it is distributed spatially and temporally, in order to understand why some species have remained monogamous, whereas others appear to have been caught up in "run-away" sexual selection.

Mound-Builders

No account of the breeding behaviour of birds in Australia is complete without a description of the megapodes or mound-builders. The thirteen species are found in Australia, New Guinea and many nearby and Pacific islands, including the Philippines and Sulawesi. Most species occupy wet forests, though the Malleefowl (*Leipoa ocellata*) occupies semi-arid woodlands in southern Australia (Frith 1985).

Female megapodes have been relieved of parental duties but make up for them by laying large numbers of eggs, up to 35 in captivity (Fleay 1937). In most species it is the male that is primarily responsible for incubation and he achieves it in a remarkable and laborious manner.

In some species eggs are simply laid in sandy soil and left. The heat of the sun, or exceptionally of volcanically warmed rock, provides a sufficient temperature for development. In dense forests large mounds are constructed of vegetation and litter, which ferment and produce heat.

The Malleefowl, although an atypical megapode, has been studied most thoroughly (Frith 1956, 1962a) (see Plate 9). Because it lives in such a dry environment fermentation of leaf litter may be slow and difficult. In the winter or late autumn the male and female Malleefowl dig a deep hole in the sand to which leaf litter is added. As this is the wetter season there is usually sufficient rain to soak the litter. When this has occurred an egg chamber of smaller items of litter is placed in the centre and all is covered with sand. The litter then starts to ferment, slowly at first, then more rapidly by late spring. By summer, the heat from the sun becomes as important as heat from fermentation.

After the female starts to lay eggs in September the male is responsible for maintaining the mound. He is apparently able to measure the temperature accurately with his tongue. If fermentation is raising the temperature too much he will remove some of the sand to allow heat to escape. When the sun is very hot a thick layer of sand may be added to shield the eggs and in autumn, when there is little heat from fermentation the layer of sand may be thinned to allow the warmth from the sun to penetrate. In this way the male can keep the egg chamber very close to 34°C for about six months. Females may lay their large eggs at about five day intervals over a long period. Eggs hatch after about 49 days depending on the temperature within the mound. Young birds dig their way out through the sand and run away from the mound, probably unseen by their parents. They are so well advanced though that they can fly within two hours. A large investment by the female into each egg and by the male during incubation has removed the need for parental care.

In dry years Malleefowl do not complete building their mounds and no eggs are laid. Booth and Seymour (1984) added 400 litres of water to two mounds in a very dry year and demonstrated that the litter needs to be wet in order to ferment and become warm. Although the mounds did increase in temperature and the males continued to work on them for longer than those with unwetted mounds, no eggs were laid. Perhaps food was insufficient for the female to form eggs.

The Orange-footed Scrubfowl *(Megapodius reinwardt)* was studied in north Queensland by Crome and Brown (1979). It apparently breeds in pairs, but more than one pair may share a mound of fermenting vegetation. The female contributes nearly as much work to mound construction as does the male, and work continues throughout the year. Although Scrubfowl are apparently able to maintain reasonably stable nest temperatures they are possibly less efficient than Malleefowl and temperatures in the nest drop markedly after heavy rain. The Scrubfowl is a widespread species among the islands to the north of Australia and displays a wide range of nesting behaviours. Detailed studies in different habitats would be most interesting.

Australian Brush-turkeys (*Alectura lathami*) are widespread in eastern Australia and remarkably tolerant of human disturbance (Plate 3). Surprisingly they have been studied little in the field. They construct their mounds of leaf litter, which may be 3–5 m across, over three to eight weeks. Sites are chosen that have a dense canopy, which provides much leaf litter and maintains a high humidity (Jones 1988). Mounds are often placed on the site of an earlier mound, though the same mound is rarely used in successive years. Mounds are associated with a dense growth of saplings, which have perhaps sprung up from seeds raked into earlier mounds. They are often close to thickets of the introduced *Lantana camara*, which perhaps provides a screen from other males. About half of the mounds receive no eggs, often because their males have been expelled by neighbouring birds which occasionally appropriate a second mound.

It is hard to explain the evolution of the unique method of incubating of the megapodes, though Frith made two suggestions. First it could be a relict of the reptilian behaviour of laying eggs in sand and leaving them, though this seems unlikely. Second, and perhaps more likely, early mound-builders could have laid their eggs in more normal ground nests, which they covered when the nest was unattended, thereby hiding the eggs from predators.

Long periods of neglect still may have led to successful hatching, at the same time allowing the parents extra time to feed.

Summary

Australian birds display a wide range of mating strategies and parental behaviour. Male lyrebirds, bowerbirds and birds of paradise perform the most spectacular displays and attract one or more mates. Although much has been learned about these birds in the last decade or two, there is still a great deal of mystery about the evolution and function of their behaviour. Only a few of the more "normal" monogamous species have been studied in detail and it is likely that, as more receive attention, further surprises will be revealed.

The unique incubation behaviour of the megapodes has been well described but we know rather little about the factors that influence hatching success and nothing about the fate of chicks after they leave the mound. It is important that these be investigated, as the Malleefowl is suffering a reduction in numbers and contraction of range. It is necessary to find out the contribution of habitat loss, fragmentation and degradation and the role of introduced predators in this decline.

7. Co-operative Breeding

Whereas in some species of birds the males have been emancipated from parental care, in others the female has assistance not only from her mate but also from additional helpers. This is termed *co-operative* or sometimes *communal breeding*. The additional members of the group, for whom the admittedly anthropomorphic term *helper* has become established, take food to the nestlings and fledglings and assist in territorial disputes. In a few species more than one male and female in the group breed in a season. Co-operative breeding is of particular interest because the helpers seem to behave in an altruistic manner. In genetic terms they are increasing the contribution that other individuals make to the next generation at the expense of their own fitness. This has been interpreted at the level of selection acting on the species or population, the kin group or the individual. There have also been attempts to relate co-operative breeding to the habitat, feeding ecology and demography of the species. Australia has a very large number of co-operative species, which are highly diverse morphologically and ecologically. Before discussing the theoretical aspects of the evolution and ecological correlates of co-operative breeding, I should like to give examples of the different systems of co-operative breeding.

Several classifications of co-operative breeders have been attempted, some of which have been very complex (Dow 1978b). I suggest that the major features are: (1) frequency of groups larger than a breeding pair, (2) group size and cohesion and (3) the interactions between members of different breeding groups. Most Australian species, at least, fall into one of the four groups outlined in Table 27. Many species typically nest as pairs, with one or two helpers in a few per cent to over 50 per cent of cases. Others live in tight groups of three to about a dozen, with a simple pair being a rare occurrence. In a few species breeding groups form larger assemblages outside the breeding season, which have some permanency from year to year. A handful of species show

Table 27. Four main types of co-operative breeders among Australian birds.

Type	Breeding Group Size	Non-breeding Group Size	Examples
Facultative	2–5	occas.> 5	Fairy-wrens, Kookaburras
Obligatory	3–12	occas.>12	Choughs, Babblers Sittellas
Two-tier	2–4	8–20	Some Thornbills
Complex	3–11 (at one nest)	20–200 (in colony)	Miners

interactions at several levels in colonies that at first sight appear almost chaotic.

Facultative Co-operative Breeders

These include perhaps the majority of species that have been recorded as breeding co-operatively, including many that have not yet received close study. The best known are the fairy-wrens (Superb Fairy-wren — *Malurus cyaneus* Rowley 1962, 1965; Nias 1984, 1986; Splendid Fairy-wren — *M. splendens* Rowley 1981; Russell and Rowley 1988; Variegated Fairy-wren — *M. lamberti* and White-winged Fairy-wren — *M. leucopterus* Tidemann 1980, 1986) (Plate 37) and the Yellow Robin (*Eopsaltria australis*) (Marchant 1985) (Plate 38). Ian Rowley's work on the Superb Fairy-wren was the first on a co-operative species that involved individually marked birds. It was to provide a benchmark for future studies of co-operative breeding.

Simple pairs of fairy-wrens are, on the whole, more common than groups of three or more, but the frequency varies between species and within species between years (Table 28). Fairy-wrens nest in low shrubs, including exotic species, laying several clutches of three or four eggs throughout a long season. They feed in open ground but retreat to shrubs when threatened by predators. In the autumn some juvenile females leave the family group and before breeding starts in the following spring there is a departure of some males and most of the remaining females. New females immigrate into the population and replace breeding females that have died. Some pairs are thus left with extra males that are typically one year old, and occasionally young females. These birds are usually offspring of one or both of the breeding birds, at least where colour-banding has allowed this possibility to be recognized. Only the breeding female builds the nest and incubates the eggs, but all

Table 28. Frequency of different-sized breeding groups in two species of fairy-wrens over several years.

Superb Fairy-wren		1♂ 1♀	2♂ 1♀	3♂ 1♀	1♂ 2♀	2♂ 2♀	5+
(Rowley 1965)	1956	11			4		
	1957	4			3		
	1958	1			5		
	1959	7			0		
	1960	6			2		
(Nias 1987)	1982	4	0	0	0	4	3
	1983	9	1	0	3	1	0
	1984	3	2	1	1	0	1
	1985	6	3	0	0	1	0
	1986	5	3	3	0	0	1
Splendid Fairy-wren							
(Rowley 1981)	1973	2	4	1	1	0	0
	1974	4	5	0	0	1	0
	1975	6	3	1	0	0	0
	1976	2	7	0	0	0	1
	1977	2	3	0	1	2	1
	1978	5	2	1	0	0	1
	1979	4	5	2	0	1	1

members of the group feed the young. The mother, and occasionally other birds perform a remarkable "rodent run" display to distract potential predators from the nest. Helpers may depart during the breeding season, or after it, to replace vacancies in neighbouring groups, resulting from the disappearance of breeding birds.

Pairs with helpers may have higher breeding success than those without helpers, though this is not always the case and its significance is hard to interpret. I shall return to this later. Fairy-wrens are unusual among co-operative breeders in being sexually dimorphic, with old males becoming blue earlier in the spring than young males.

Obligatory Co-operative Breeders

Of the species that almost invariably breed in groups of three or more the White-winged Chough (*Corcorax melanorhamphus*) (Rowley 1978) and Grey-crowned Babbler (*Pomatostomus temporalis*) (Counsilman 1980; King 1980) (Plate 39) are the best studied. Both belong to the corvid family in its widest sense, though the White-winged Chough is not closely related to the

Eurasian choughs (*Pyrrocorax*) and the Australian babblers are totally unrelated to the Old World Babblers despite remarkably similar behaviour (Gaston 1977).

There are many similarities between the way of life of the Chough and Babbler, and I shall discuss them together. Both are rather large birds with curved beaks, which probe and rake up leaf litter for large invertebrates. They feed as a tight group, but being weak fliers retreat to bushes and trees when danger threatens. The sexes are indistinguishable but the age of a bird can be assessed by the colour of its iris. Babblers attain bright yellow eyes by the age of about two and Choughs' eyes become redder until they are four or five years old.

Babblers live in groups of typically three to six and Choughs usually of four to eight, though simple pairs are not unknown. Both species can be noisy and excitable, but show little aggression within the groups though they may dispute neighbouring territorial boundaries. Sex ratio does not appear to differ greatly from unity and females do not disperse more than males. In fact most immatures tend to stay with their parental group unless for some reason it disintegrates. All group members contribute to nest building, though young Choughs are inexpert at this. Babblers build untidy nests for breeding and roosting and Choughs construct mud nests on a horizontal branch. There is strong circumstantial evidence that two or more female Choughs may lay in a single nest. This is usually termed communal breeding rather than or as well as co-operative breeding. All birds feed the young in the nest and after they have fledged. Although young in larger groups do not receive more food, members of larger groups need to work less hard to rear offspring. Larger groups tend to rear more young to independence each year and juveniles survive better in their first year in larger groups, at least in Choughs (Table 29). Choughs are very long-lived once they have become breeding birds.

Two-tier Systems

Harry Bell discovered that Buff-rumped (*Acanthiza reguloides*) and Striated Thornbills (*A. lineata*) show a predictable change in group size between breeding and non-breeding seasons (Bell 1983a, 1985b; Bell and Ford 1986). The story is most complete for the Buff-rumped Thornbill, a tiny insectivorous bird that forages from the ground, bark and foliage in eucalypt woodlands and open forests. From late summer to late winter they roam around in groups of a dozen or more birds, which Bell called "clans", often feeding on the ground with other species. As spring

Table 29. Breeding success of White-winged Choughs in different sized groups (from Rowley 1978, Table 10 and Figure 5).

Group Size	Fledglings per nest	Number of nests	Percent fledglings surviving one year
2	0	2	—
3	0.18	11	—
4	1.11	35	39%
5	1.30	44	29%
6	1.00	33	58%
7	1.04	23	70%
8+	1.50	38	43%

approaches there is increased aggressive activity within the clan, which splits into pairs or groups of three or four birds. The four or five groups of a clan occupy sections of the overall clan territory (Figure 29). Each group attempts to breed, often several times. About 80% of nests fail and by early summer unsuccessful groups have joined more successful groups again. All group members help to feed their own nestlings and fledglings and as the clan re-forms unsuccessful birds will feed the fledglings of other groups.

Immature females may emigrate to other clans in autumn or late winter, whereas males tend to remain in the clan and unless they gain a breeding position, help any surviving parent. Both clans and groups therefore show a similar constitution from year to year.

In many ways Buff-rumped Thornbills, and probably most of the thornbills, behave like the fairy-wrens. They differ in not being sexually dimorphic, and in belonging to a clan as well as a group, which can also be co-operative.

Complex Systems

The miners, which are large honeyeaters that feed mostly on insects and exudates, present the most complicated of avian social organizations. The Noisy Miner (*Manorina melanocephala*), which occupies open woodland has been studied near Brisbane by Douglas Dow (Dow 1970, 1978a) whereas the Bell Miner (*M. melanophrys*), which occupies wet gullies has been investigated by a succession of observers near Melbourne (Swainson 1970; Smith and Robertson 1978; Clarke 1984, 1988).

Both live in colonies, which may reach 100 birds, usually surrounded by unoccupied habitat. These sites have abundant

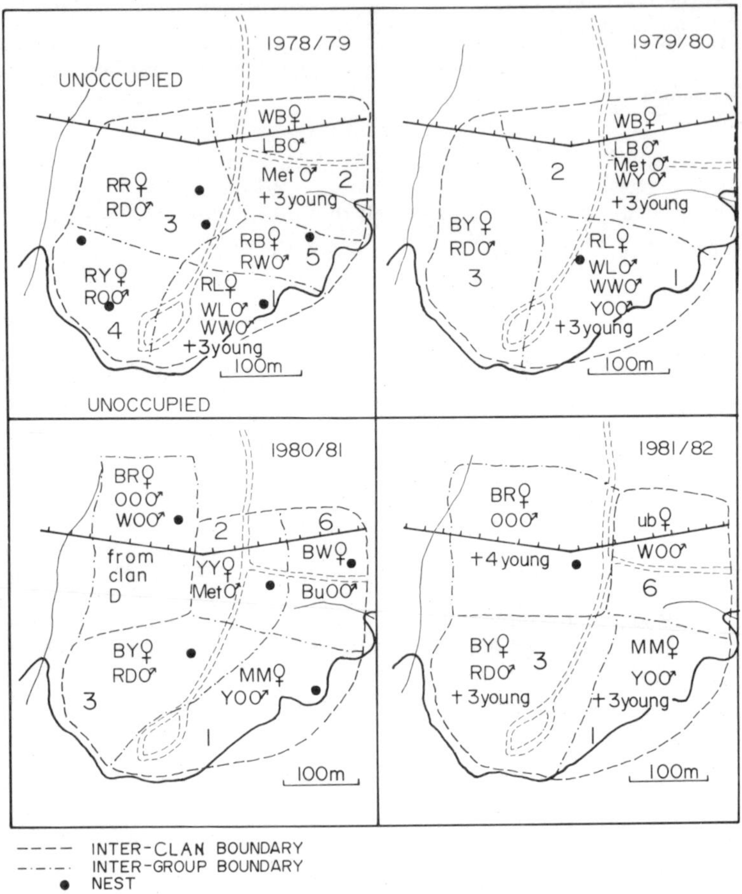

Fig. 29. Clan and group territories in Buff-rumped Thornbills (copied from Bell and Ford 1986 with permission of publisher).

food such as psyllid lerps or other insects. Such rich areas are defended vigorously by the miners (see Chapter 4). Most other birds are driven out of the area and may even be killed. Predatory birds, as well as mammals and reptiles, may also be attacked. Colonies consist of several smaller groups called "coteries" by Dow and "clans" by Clarke. The sex ratio is highly imbalanced with three males to each female. The female alone builds the nest and incubates the eggs but the other clan members feed the young. Several males may mate with an individual female in Noisy Miners and perhaps males mate with more than one female. Certainly males will feed at a succession of nests, belonging to several females throughout the long breeding season. They spend most time visiting young of females within their coterie or clan. Indeed the effort that Bell Miners put into feeding at a nest is directly correlated with how closely related they are to the young (Clarke 1984). Nests may receive visits from ten or more adults at rates reaching 80 per hour. Despite all this attention, breeding success is very low, and it is possible that some visitors do unintentional or even wilful damage.

Miner colonies have not been studied for long periods, but it appears that there is little interchange between neighbouring colonies as intruding birds are attacked and driven out.

Although many aspects of the miners' behaviour are undoubtedly co-operative, they can at the same time be fiercely competitive. As regards the breeding system the Noisy Miner is very likely polyandrous as well as perhaps polygynous.

What are the Advantages of Co-operative Breeding?

With co-operative breeding in birds spanning such a variety of behaviours it is not surprising that there have been many explanations of its evolutionary and ecological significance. Many of these are by no means mutually exclusive and indeed very different explanations could be correct in different species or circumstances.

With regard to evolution, benefits could accrue to the species or population by group selection, to the individual directly by increased survival and lifetime reproductive success or indirectly through the individual's relatives. I shall consider these possibilities separately.

Group Selection

Skutch (1935), working in central America, first described co-operative breeding in a paper entitled "Helpers at the nest". His interpretation was couched in terms of the good of the species, which was how behaviour was usually explained at that time. In particular, he felt that some individuals would forgo the opportunity to reproduce so that the birth rate of the population could be kept modest and the risks of overpopulation minimized. (He still holds this view — Skutch 1987.) Rowley (1965) also initially interpreted the behaviour of fairy-wrens at the species level, with helpers allowing the species to respond to a widely varying climate.

Most behavioural ecologists now regard selection as acting on individuals. Those with more progeny will contribute the most genes to the next generation. Birds that help others to breed but do not breed themselves would be expected to contribute fewer, or even no, genes to the next generation. Helping therefore ought to be a less profitable strategy than breeding and any genes that contribute to helping behaviour should be progressively lost from the population. In fact one of the weaknesses of interpretations of co-operative breeding is that we have absolutely no information on the genetic basis for helping.

Group selection requires rather precise conditions to prevail in the face of individual selection. Populations need to consist of small groups with rather limited gene flow between them. Wynne-Edwards (1986) believes that many vertebrate populations are indeed like this. Some co-operative breeders consist of particularly isolated units. There is as yet no concrete evidence for the existence of group selection in co-operatively breeding birds.

Individual Selection

There are two separate parties in a co-operatively breeding species, the helpers and the breeders. The advantages to the breeders may seem fairly obvious, but they could be counter-balanced by disadvantages such as competition. These will be discussed later. For the helpers potential advantages fall into two groups depending on what one sees as the alternatives that are available.

For a start, helpers may have chosen to forgo the option of breeding. Alternatively, this option may not have been available and helping would merely be an alternative to wandering without a territory or occupying a territory with no mate.

Successful breeding may require experience in some species. Young Choughs have little idea of how to build a mud nest (Rowley 1978). Attempting to breed may require leaving the family group with its familiar territory, collective knowledge of where food can be found and the advantages of vigilance and protection from predators or other aggressive species (Heinsohn 1987). In some birds, particularly those that are long-lived, a few years' apprenticeship in the safety of the parental group may lead to a more productive lifetime than an early departure to the precarious life in the outside world. Rowley draws an analogy with the seabirds that delay breeding for several years (see Chapter 5).

A rather separate argument, particularly developed by Emlen (1978) and Emlen and Vehrencamp (1985) proposed that helpers simply do not have the option of breeding. There are ecological constraints preventing them from doing so. If given the opportunity they would breed. The constraints fall into three main groups, (1) a shortage of food, (2) a lack of territories and (3) no mate. Birds need extra food to be able to breed and in poor years this food may not be available. Fairy-wrens and other birds may fail to breed in some years even though they have mates and territories (Nias 1987). Where a scarcity of food is compounded by inexperience, breeding may become even more out of the question. Emlen (1978) found that African bee-eater (*Merops bulockoides*) colonies have a far higher proportion of helpers in years when insects are scarce. In aseasonal environments where food at the best time of year may not be much more plentiful than at the worst time, it may always be difficult to breed.

A major requirement for successful breeding is a good territory, which includes safe breeding sites and refuges from predators as well as sufficient food. Brown (1978, 1987), in particular, has developed the argument that co-operative breeding is found in birds that have rather specific habitat requirements, and that tend to be long-lived. They may occupy rather equable environments where food is rarely scarce so that breeding vacancies only arise infrequently. Young birds may have to wait their turn to breed, often for several years. An excellent example of a bird living in these conditions is the Florida Scrub Jay (*Aphelocoma coerelescens coerelescens*) (Woolfenden and Fitzpatrick 1984). A long term study on this bird has shown that remaining in the parental territory is a good strategy, at least for males. You can keep a lookout for vacancies in neighbouring territories as well as having a reasonable chance of inheriting your father's property, plus his mate and other helpers (which you may have helped to rear).

The shortage of mates is another reason why breeding may be impossible. The evidence for this being a factor in co-operative breeding is less convincing, however. In many species, as in the fairy-wrens, young females tend to disperse whereas young males tend to stay at home. Young females probably suffer high mortality in their solitary wanderings and adult females may also survive less well than males. This leads to a shortage of females and a surplus of males. Emlen (1978) used Rowley's data to show that the imbalance of the sex ratio (in favour of males) was correlated with the proportion of groups with helpers. He used this to support his contention that females were a limiting resource for males, leading to co-operative breeding. However, as most helpers in fairy-wrens are males such a correlation is inevitable. It does appear that a shortage of females can occasionally prevent males from breeding even when suitable territory is available (Nias 1987). The presence of helper females in some fairy-wren populations (Rowley 1981; Nias 1987) suggests that a shortage of mates is by no means the only, or even main, cause of co-operative breeding. Unfortunately very few co-operative breeders are sexually dimorphic, so that the shortage of mates is hard to assess, though male-biased sex ratios do appear to be common.

The ecological constraints model of Emlen, based on the studies and theorizing of Brown and many others, comes close to a comprehensive explanation for the evolution of co-operative breeding. However, it does not answer the question of why birds that remain in the parental territory should actually help in their breeding attempts. There are two possible explanations: (1) the parents may benefit from help, and so will the helper indirectly if related to the parent; or (2) the helper may benefit from being in the territory, but has to help to appease the breeder, which might otherwise drive away a potential competitor (Emlen 1982b).

Kin Selection

Genes may be contributed to the next generation through offspring, but identical copies can also pass on through relatives. Individuals share about half of their genes with their parents, offspring and siblings, and about one quarter with grand-children, nephews, nieces and half-siblings. The component of selection that involves indirect transfer of genes from relatives to the next generation is known as *kin selection*. Where individual and kin selection are combined the term *inclusive fitness* is used. This represents the overall genetic contribution of an individual but is a hard concept to appreciate and measure.

Kin selection could be a major reason for co-operative breeding if the indirect gain in fitness exceeds any loss in individual fitness (Brown 1978, 1980). For this to be the case there are two requirements. The helper needs to be related to the breeding bird and the young that it assists in rearing. Second, the helper needs to improve the breeding success of the breeders.

In almost all cases where individually marked birds have been observed, helpers have been found to be the offspring, or siblings, of one or both of the breeding birds. Typically they are young birds that have been retained from previous breeding seasons. There are cases where unrelated birds join groups that they help, but usually only after the disintegration of their own parental group. On average, helpers are related to the nestlings they help, in the case of Splendid Fairy-wrens by 0.47 and in Florida Scrub Jays by 0.40 (Russell and Rowley 1988; Woolfenden and Fitzpatrick 1984). The scene is set for kin selection.

Do helpers improve the success of the breeding birds? Rowley's Superb Fairy-wrens are often quoted as an example of the advantages of having helpers. The exact figures though need to be examined closely. In fact groups rear more young per nest and per egg than do pairs mainly because more of their eggs hatch (Table 30). This is a curious finding as help is not given until after

Table 30. Relative success of groups and pairs of Superb Fairy-wrens (from Rowley 1965).

	Groups	Pairs
Number of nests	32	32
Eggs laid	97	96
Eggs hatched	76*	52
Nestlings fledged	69	39
Independent fledglings	34	24

*Proportion significantly different (X^2, $P<0.01$).
No other differences are significant.

hatching. Perhaps groups live in better territories, or perhaps some females are better than others at choosing safe nest sites. This could be why they have helpers in the first place, because they reared young the previous year. Nias (1986) found that the quality of a territory (especially good nest sites) occupied by Superb Fairy-wrens near Armidale had more influence on how many young were reared per season than did the number of helpers attending the nest.

Helpers can also assist breeders by decreasing their workload, as is found in the Babblers (Table 31). They can also look after recently fledged young and allow the mother to re-nest more rapidly. When all these factors are considered together it is clear that helpers can increase the parents', in particular the mother's, chances of survival. Russell and Rowley (1988) found that female Splendid Fairy-wrens with helpers were more likely to be alive to

Table 31. Effects of helpers on total feeding rate and parental effort in Grey-crowned Babblers (from Brown *et al.* 1978, Table 1 and Figure 1).

Number of Helpers	No. feeds+ per hour	Percent of visits by father*	mother*
0	6.15	45%	55%
1	6.65	35%	34%
2	7.54	22%	31%
4	4.88	10%	21%
8	8.33	22%	18%

+not significantly correlated with number of helpers.
* both showed negative correlation between *number* of visits and group size.

breed in the following year. Their life-time reproductive success could be effectively doubled in this way.

Reviews of the relationship between group size and reproductive success have shown that in most cases they are positively correlated (e.g., Koenig 1981). The advantages of large group size may extend well beyond the fledging stage as is found in Choughs (Table 29). It is harder to be sure how much of the increased success is due to the direct effects of the helpers, though this appears to be the case in Babblers (Brown *et al.* 1982). Despite the circumstantial nature of much of the evidence, it seems likely that on average helpers do increase the breeding birds' fitness, and indirectly their own, in the majority of cases. What is not at all clear is whether this indirect component of fitness is simply a bonus above the individual advantages of delaying breeding and dispersal, or whether it is a major factor that on balance counteracts the disadvantages of not breeding. In the most complete assessment of kin selection available, Woolfenden and Fitzpatrick (1984) suggested that Florida Scrub Jays may increase their fitness indirectly by helping, but that this benefit is small compared with the advantages of staying within the parental territory until a vacancy appeared. Given a breeding opportunity it would certainly be advantageous for Jays to take it. In Bell Miners kin selection could well be a major component, as there was a direct correlation between relationships and feeding rate (Clarke 1984).

In summary, co-operatively breeding birds may be sufficiently long-lived that delayed breeding may be advantageous for similar reasons to those occurring in most seabirds.

More often though the opportunity to breed may not be available for the first year of life, even though young birds would be willing and capable of reproducing. They can make the best of circumstances by awaiting vacancies in breeding territories, a potential mate or better conditions from within the safety of a proven territory, while at the same time gaining some experience. As they are usually helping relatives any increase in the success of breeding birds will provide an increment in their own inclusive fitness. In exceptional circumstances this increment may be larger than they could attain by attempting to breed themselves.

Ecology and Co-operative Breeding

Why do some species of birds breed co-operatively whereas their close relatives do not? Do particular environments provide conditions that are particularly favourable to the evolution of co-operative breeding? Why has Australia so many co-operative species? Attempts have been made to answer these questions both in a global and an Australian context. Some have warned though that co-operative breeders are such a diverse collection of species that we should not expect simple answers (Fry 1977; Dow 1980).

Brown (1978, 1987) proposed that co-operative species mostly share common demographic features. They tend to be sedentary, territorial and long-lived. This means that relatives tend to stay together, allowing co-operation through kin selection. It is interesting that the few migrants that are co-operative are aerial feeders that migrate diurnally (bee-eaters, hirundines, wood-swallows). Presumably families can keep together. There is strong competition for breeding territories that provide safe nest sites, refuges from predators and good feeding areas. As the occupants are long-lived, vacancies in these territories are infrequent, and there is less opportunity for young birds to start breeding. Individuals that delay dispersal and breeding are favoured over those that attempt to breed in poor quality habitat in their first year. Brown's own studies on American jays, as well as those on Florida Scrub Jays (Woolfenden and Fitzpatrick 1984) provide the best evidence for this "habitat saturation" hypothesis.

To extend this demographic pattern to a species' ecology, it is clear that a reasonably stable environment is required. Stability is usually meant in terms of low seasonality of resources, particularly food. Rowley (1974) noted that Fairy-wrens do not experience a harsh winter, with food shortages, as species at higher latitudes do. As competition for food will be less severe, breeding birds can permit their young to remain in the parental territory for long periods. A further important point is that Australian environments probably do not experience a regulaı superabundance of food in spring. The lack of seasonality in some Australian environments may mean that breeding is difficult and not very productive, while survival is relatively high. Again the urgency to leave and breed is removed, at least until young birds have gained experience. Woinarski (1985b) linked the high incidence of co-operative breeding in Australia with long breeding seasons and low clutch sizes, as a response to an aseasonal environment (see Chapter 5).

There are differences in the frequency of co-operatively breeding species in different parts of Australia. Dow (1980) showed that they tended to be concentrated at intermediate latitudes and nearer the coast than the centre (Fig. 30a). There

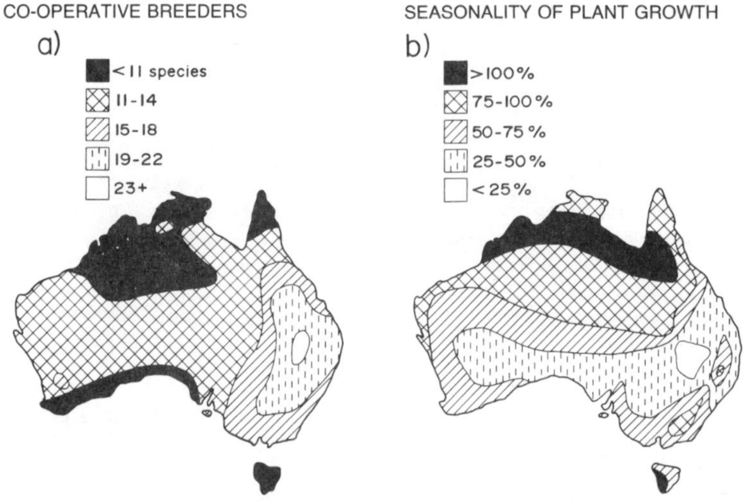

Fig. 30. (a) Frequency of co-operative breeders in different parts of Australia (b) index of seasonality in different parts of Australia (copied from Ford, Bell, Nias and Noske 1988 with permission of publisher).

appear to be rather few species in the seasonal cool temperate south and the monsoonal north. To some extent this is due to the paucity of studies, at least in the north and centre. However, the resemblance between Dow's map and that of Nix (1976) on seasonality of resources is remarkable (Fig. 30b). Co-operative breeders appear to peak just west of the Great Dividing Range in New South Wales and southern Queensland, the region of lowest seasonality of resources.

If one looks at habitats, a further interesting pattern emerges (Table 32). Co-operative breeders appear to be more common in

Table 32. Numbers of co-operatively breeding species in each of several habitats in Australlia (from Ford, Bell, Nias and Noske 1988).

	Rainforest	Eucalypt forest	Eucalypt woodland	Semi-arid woodland	Desert
Number of species	7	35	48	36	9
Proportion of all species	7%	27%	31%	33%	15%

eucalypt woodland than in both drier and wetter habitats. Woinarski and Cullen (1984) showed that arthropods on foliage are in fact less seasonal in abundance in drier eucalypt woodland than in wetter forests. This is supported by the fact that several forest birds (e.g., Pied Currawong *Strepera graculina*, Flame Robin *Petroica phoenicea*) tend to move inland into drier habitats in winter. Even rainforests appear to display a marked seasonal change in the abundance of arthropods (Lowman 1982).

Another feature that has been proposed as contributing to co-operative breeding is an arid (Harrison 1969) or unpredictable climate (Rowley 1968, 1976). Aridity *per se* is perhaps not important, but the fact that dry years, when breeding is difficult, alternate with good years when breeding is easier, ensures that in some years a significant proportion of birds do not breed. Emlen (1982a) presented the hypotheses of stability and unpredictability as alternatives, but in fact as they refer to different time-frames they can both be important. Habitats without regular seasonal changes, but where poor years follow good ones, would be those in which a surplus of adult birds over the number of breeding opportunities could arise. These adults, prevented from breeding, could then assist others that are breeding.

The tendency for co-operative breeders to be found in habitats of intermediate climate (semi-arid) has also been noted in Africa (Grimes 1976) and India (Gaston 1978). In North America the few co-operative species are mostly found in evergreen oak woodland or scrub.

Given then that aseasonal but rather unpredictable habitats may lead to strong competition for breeding territories why are only some species co-operative breeders? I have separated the bird species occupying eucalypt woodland near Armidale into co-operative and non-cooperative species and compared them in terms of diet, foraging substrate and foraging method (Table 33).

Table 33. Foraging ecology of cooperative and non-cooperative breeding species in eucalypt woodland (from Ford, Bell, Nias and Noske 1988).

Diet					
	Seeds, fruit	Nectar	Arthropods		
Co-operative	0	4	20		
Non-cooperative	6	5	29		
Foraging Substrate					
	Ground	Leaves	Bark	Air	
Co-operative	10.5*	7	3.5	2	
Non-cooperative	6.5	13	3	9.5	
Foraging Method					
	Glean	Snatch	Pounce	Hawk	Aerial Pursuit
Co-operative	19	0	2	0	2
Non-Cooperative	13	7	3	5	4

*Some species that use two foraging substrates scored 0.5 for each.

Co-operative breeders are mostly dependent on arthropods, a feature previously noted elsewhere by Grimes (1976), Brown (1978) and Gaston (1985). Some honeyeaters are co-operative but these are among the least nectarivorous members of the family. As noted previously foliage arthropods, at least, are not markedly seasonal in eucalypt woodlands, whereas both nectar and seeds probably show peaks in abundance (e.g., Ford 1979 for nectar). It is also interesting that seed-eaters do not show the reduced clutch sizes that insectivores do in Australia (Table 22), which may be because seed crops are seasonal in their abundance.

Among birds that feed on arthropods, those that are co-operative breeders tend to be ground foragers. This trend has also been noted on other continents (Gaston 1985). A few data indicate that arthropods on the ground are less seasonal in abundance than those above the ground (Huddy 1979). Some species feed far more on the ground in winter than in other seasons (Scarlet Robin – *Petroica multicolor* – Huddy 1979, Buff-rumped Thornbill – Bell 1985a). Therefore, the question of availability and seasonal abundance of food in different sites should be examined more closely.

Co-operative breeders typically pursue their prey by gleaning or aerial pursuit rather than by employing sit-and-wait tactics like hawking, snatching and pouncing. Here again the larger food taken by the sit-and-wait predators may vary more seasonally. This is supported by the fact that many of these species are migratory, presumably leaving in winter when food is scarce.

If lack of seasonality in resources and irregularity from year to year can lead to an excess of potential breeders over suitable territories, why should non-breeding birds seek vacancies from within the parental territory? I believe that the main answer to this is safety from predation. Predation, of nests or free-living birds, is a major cause of mortality in co-operative breeders. Many species actually build domed or partly enclosed nests. The fairy-wrens show a rodent-run distraction display (Rowley 1962). Living in groups means that predators are more likely to be detected and avoided and less time need be wasted in looking out for them (Heinsohn 1987). In some cases they can even be attacked or repelled. Miners collectively can remove far larger birds from their colonies (Dow 1977). Choughs can also defend themselves in a group from Magpies (*Gymnorhina tibicen*), which can hound solitary birds (Rowley 1978).

Birds in moderately open habitats probably have the most to gain from living in groups. Individuals that detect predators have enough time to warn other members of the group, intentionally or otherwise, to allow them to escape to cover. In dense habitats like rainforests once a predator has been detected it is often too late to escape or warn others. In open habitats like deserts there may be little cover, though collective manoeuvres in flight may deter the predator. Ground foragers may be particularly vulnerable to predation in moderately open habitats, and rapidly retreat to shrubs or trees when danger threatens. Finally, those that forage by gleaning are seeking prey that are close to them, whereas sit-and-wait foragers look for prey several metres away. The former may be less aware of predators lurking a few metres away. Sit-and-wait feeders also take more active prey, and may be better at detecting a moving predator or may have greater manoeuvreability in escaping predators.

Summary

Co-operative breeding is practised by many Australian birds.

There is no shortage of hypotheses to explain its function and the ecological reasons for its evolution. Evidence has been provided from several Australian species to support each of these hypotheses. True tests, though, can probably only be carried out by comparisons of closely related co-operative and non-cooperative species in the same area. The idea that the aseasonal but unpredictable nature of the Australian environment has led to the high incidence of co-operative breeding needs to be examined and refined.

8. Population Ecology

Population Ecology is the study of the factors that influence the size or density of a population in an area. Typically, differences in density between areas or over time are investigated. Such factors may be *independent of density* in that they tend to kill a fixed proportion of the population or limit reproductive rate to a particular level for the species. They may also be *dependent on density*. Greater numbers may mean a disproportionate mortality, if for instance, a particular resource is in short supply. Breeding success may decline at high densities though it may also be lower if densities are very low, as mates may be hard to find, or there may be a lack of social stimulation from other breeding pairs.

Estimating density

One of the first problems in population ecology is to count or estimate the numbers of individuals or density of the population. Where one has animals as mobile as birds this can be no easy matter. It can be made even harder by large scale migration or nomadism, and by seasonal changes in foraging behaviour and the incidence of singing and calling. It is not surprising that most counts are made during the breeding season when most birds are tied down to a nest.

Waterbirds and waders are often restricted to particular locations and especially during high tides or dry periods their numbers can be counted reasonably easily (see Jaensch 1984; Minton and Lane 1984). Sometimes their vast numbers and dense flocking can provide problems even when they are very localized. Seabirds in the breeding season can be relatively easy to count if they nest on the surface or on cliffs.

Most landbirds tend to inhabit large belts of rather homogeneous vegetation, and except for large species in open habitats, such as cockatoos, are impossible to count simultaneously. One needs to sample small areas and estimate densities within them. It is difficult to be sure that representative areas have been chosen, but a wide variety of methods is available to census landbirds. A range of these is given in Table 34. Each has certain advantages and disadvantages and the most appropriate method will depend on the species being studied and the questions being asked. Point counts are easiest to carry out but they provide little more than a rough index of abundance. Usually only small areas can be covered meaning that most species will be represented by only one or two observations. They may be adequate though if all that is required is an indication of seasonal changes or differences in abundance between habitats in common species. Such differences in abundance can be related to abundance of resources or other bird species.

Transect counts have been popular, especially in Scandinavia and have been used in Australia since 1919 (Cleland 1919). They provide the advantage of covering a large area and recording many birds of a range of species. Long transects often cover a diversity of habitats which may or may not be desirable, depending on the questions asked. Transects yield an estimate of the number of birds per unit distance, which can be converted to a density if a fixed width is used (e.g., Bell 1965). A refinement is to note the distance of a bird from the centre of the transect and vary the transect width for different species or habitats (Emlen 1971; Bell and Ferrier 1985). To some extent this corrects the problem of different species differing in their conspicuousness. The problem of birds fleeing from or being attracted to the observer remains. Transects then can produce figures for density but these tend to be underestimates (see Bell and Ferrier 1985), and it may be inappropriate to compare different habitats (Harden *et al.* 1986).

Some observers have attempted to count birds within a prescribed area. This can be reliable if the area is restricted, for instance a wood surrounded by farmland. In continuous habitat it can be difficult because much time can be spent in marking out and keeping within the area. An improvement on area-searching is to map the presence of birds over several visits and compare maps. This is particularly appropriate in the breeding season where singing males, nests or family parties reveal breeding territories. This method is used successfully in Britain by the British Trust for Ornithology. There are problems with this method in Australia though as many species breed co-operatively, breeding seasons are often long and quite a few species are either not territorial or proclaim their territories rather infrequently. Mapping requires many visits and tends to overlook non-breeding

Table 34. An outline of the most frequently used census methods (from Recher 1984; Smith 1986).

	Season	Time Required	Accuracy	Suitable Bird Types
Mapping (including nests)	Breeding	Moderate-long	Generally good	Sedentary, territorial, vocal, non-cooperative species. Need to be visibly conspicuous.
Transects	All of year	Moderate	Moderate with standardization	Most species, except inconspicuous ones.
Point Counts	All of year	Quick	Poor	Abundant species, species in dense habitat.
Capture–recapture	All of year	Long	Can be good	Species that can be captured easily but which are sedentary. Inconspicuous species.
Individual marking (colour-banding, wing-flagging etc.)	All of year, especially breeding season	Long	Good	Residents or breeding visitors, that are not too inconspicuous. Co-operative breeders. Polygamous species.

and transient birds. It is still a good method though for vocal or relatively uncommon species.

A totally different approach is to use one of the many capture-recapture methods that have been applied to other types of animals. (See population ecology texts such as Southwood 1966 and Begon 1979 for details). Many birds can be captured in mist-nets, though canopy-foraging species and large and very small birds may require different methods. Most capture-recapture methods require that a rather high proportion of the population has been marked and high immigration and emigration rates introduce difficulties. They also assume that individuals are equally catchable, and birds can certainly become wise to mist-nets. To be reliable they can be very time-consuming, but are reasonably accurate if appropriate corrections are made (e.g., see Kikkawa 1962). For individual species unique marking such as colour-banding or wing-tagging can allow resighting in the field without the need for recapture (Buckland *et al.* 1983). These methods can be difficult in cryptic birds or those with short legs. Wing tags can increase mortality and colour bands can be lost or can fade.

The particular difficulties in censusing birds in Australia were first discussed by Dwyer (1972); and Recher (1984) and Smith (1986) provided recent summaries of the relative advantages of the different censusing methods. Shields and Recher (1984), Bell and Ferrier (1985), and Ford *et al.* (1985) compared different methods in the field. When gross differences are being sought over time or space the quicker methods such as point counts may be adequate. For an idea of the composition of a community, transects or mapping are probably preferable. When one wants to know densities of individual species then mapping along with capture and individual marking are usually necessary. In any attempts to census or estimate density it is obvious that the particular questions need to be framed clearly before a method is chosen.

Population Sizes

Total population sizes are known reasonably accurately for some species, typically those that are localized. Abbott's Booby (*Sula abbotti*) breeds only on Christmas Island in the Indian Ocean, where there is an estimated total of 1,136 breeding pairs (Nelson and Powell 1986). There are only 500–1,000 individuals of the Australian subspecies of Gould's Petrel (*Pterodroma leucoptera*), breeding on Cabbage Tree Island, off New South Wales (Fullagar 1976). At the opposite extreme there are estimated to be 15 to 16 million Short-tailed Shearwaters (*Puffinus tenuirostris*) around Tasmania alone, in the breeding season (Naarding 1981).

The total populations of Wry-billed Plovers (*Anarhynchus frontalis*) and Double-banded Plovers (*Charadrius bicinctus*) are about 3,800 and 12,500 birds, all breeding in New Zealand (Sagar 1986).

Total populations of land-birds are much harder to estimate, unless they are very rare. There were only 55 pairs of Noisy Scrub-birds (*Atrichornis clamosus*) in 1985 though their numbers have increased slightly since then (G. Smith 1985). Only 100–140 Orange-bellied Parrots (*Neophema chrysogaster*) remain (Brown and Wilson 1982). However, the numbers of very common species, such as Galahs (*Cacatua roseicapilla*) or Willie Wagtails (*Rhipidura leucophrys*) are not known.

Life Tables

An alternative to censusing is to estimate birth and death rates for a population (*natality* and *mortality*). This can usually only be done for a small part of a population or an isolated one. Life tables can be constructed showing numbers or proportions of birds added and those dying each year. Rates of increase or decrease can be calculated from these.

Birth rates can be calculated by recording the numbers of nests per breeding pair or group per season, the number of eggs in each and the proportion that survive to fledging. The most valuable figure is the number of new birds produced per pair that join the breeding population the following year (*recruitment*). This is often difficult to calculate as young birds frequently disperse from their natal areas. In birds, such as many seabirds, that delay breeding until they are two to nine years of age, one needs to know the survival rate of the young birds through these early years. The mortality rate usually refers to the proportion of adult birds that die from one breeding season to the next.

Immigration and emigration rates also affect population densities as there may be a net inflow or outflow of individuals. In stable populations mortality rates and emigration rates should balance recruitment and immigration rates, at least on average. Table 35 provides an example of a life table for the Superb Fairy-wren (*Malurus cyaneus*). This population was declining slightly but annual breeding success varied from 1.09–3.60 fledglings/pair and adult mortality varied from year to year. Density therefore

Table 35. Life table for Superb Fairy-wrens at Eastwood State Forest near Armidale (Nias 1987). Survival during first year is an underestimate, as losses include emigrants.

Mean Clutch Size	3.35
Proportion of eggs leading to fledglings	0.32
Number of breeding attempts/group/season	2.52
Number of young fledged/group/season	2.70
Survival of young from fledging to next September	44%
Number of recruits/group	1.20
Adult mortality/year	55%
Adults lost per group per year	1.39

went up in some years and down in others. In practice it is often difficult to separate mortality from emigration, as birds simply disappear.

Population Changes

Populations change through the year, with increases in the breeding season and reductions in other seasons. There is also migration, post-breeding dispersal and local wandering to take into account. Palaearctic waders are in Australia from September to April, though a few non-breeders remain through the southern winter. Several species of passerines are regular summer visitors to the forests of southeastern Australia. These migrants will be discussed in the next chapter.

Some changes in density are not as regular, but occur in response to changes in the local abundance or scarcity of food. Honeyeaters have long been known to move into areas where flowering is prolific (Keast 1968b). Some more detailed studies have shown a relationship between abundance of flowers and birds (Figure 31; Ford 1983; Ford and Pursey 1982). Others though have found that correspondence between birds and flowers is less clear. This may be because other foods such as lerp and manna are being eaten when nectar is scarce (Paton 1980) or because many of the birds are transient, perhaps moving between other good patches of flowering (Pyke 1983).

Changes in breeding populations between years have been studied best. Some populations remain at remarkably constant levels over many years, whereas others fluctuate in a regular or irregular fashion and others again decline or increase steadily or dramatically. Major questions in ecology are what limits popula-

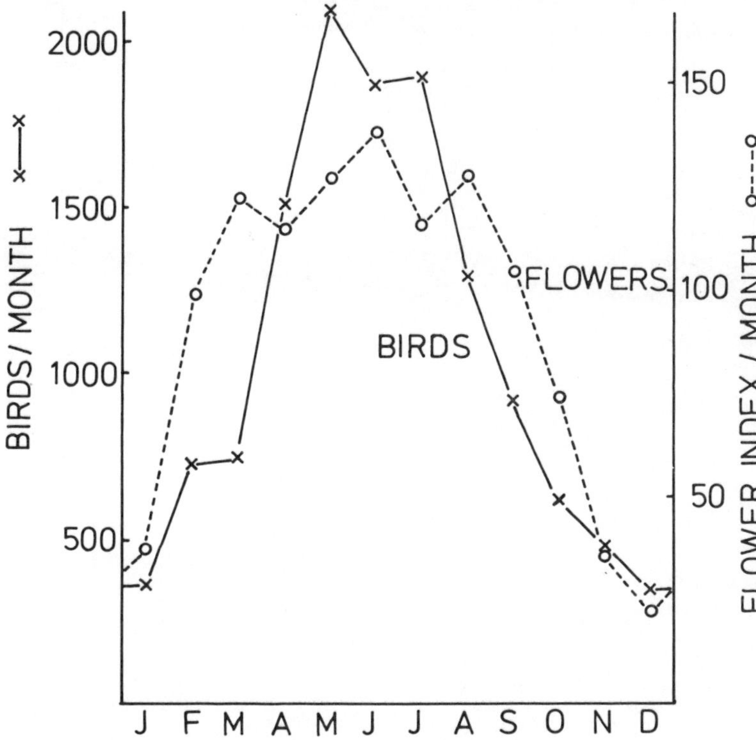

Fig. 31. Relationship between abundance of flowers and honeyeaters in sites near Adelaide, South Australia (from Ford 1983).

tion sizes and if, and how, populations may be regulated around a particular level. Before presenting examples of population changes and regulation from the Australian region I shall briefly outline the main theories on population control.

Regulation of populations

A significant part of the early theoretical development of population regulation was contributed by the Australian entomologists Nicholson (1933) and Andrewartha and Birch (1954, 1985). Nicholson believed that at least ultimately most populations were regulated in a density-dependent manner. As density rose there was a greater probability of starvation or over-exploitation of resources, or predators and parasites killed a disproportionate number of individuals. Birth rates would decline with increasing population density, while mortality rates would rise (Figure 32a).

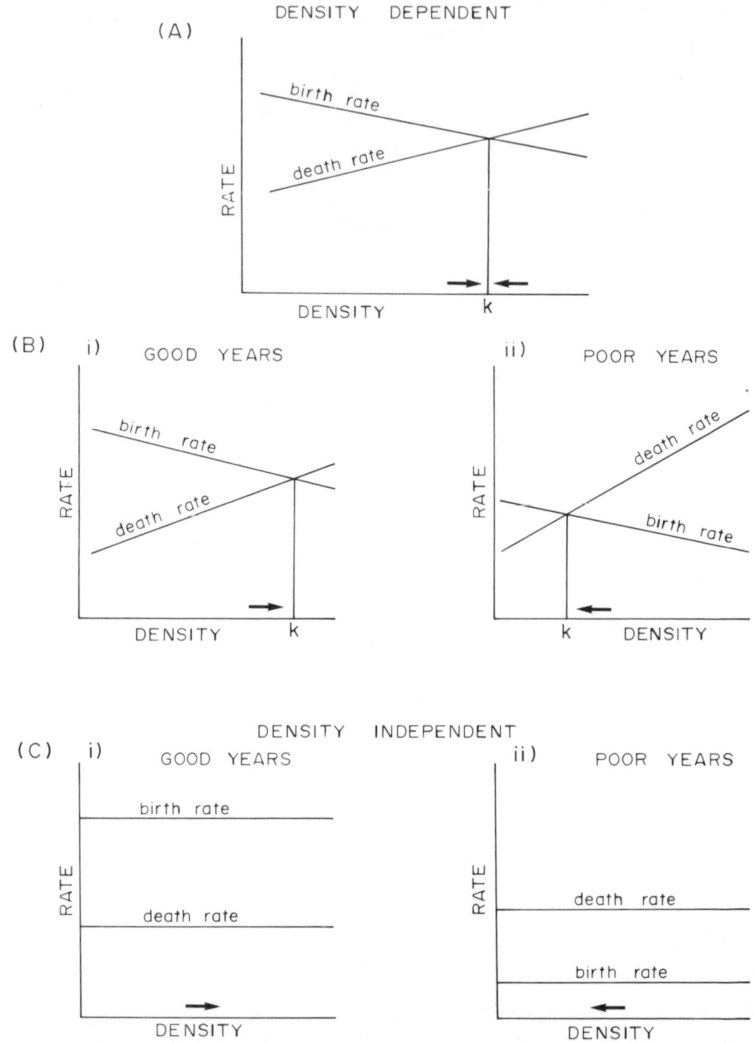

Fig. 32. Control of population density: (a) density-dependent model with birth rates falling and death rates rising with increasing density. Density tends towards a carrying capacity, K, set by the environment. (b) density-dependent model but with a fluctuating carrying capacity. Populations rise in good years and decline in poor ones so that they fluctuate around rather fixed limits rarely breaking out or going extinct. (c) density-independent model, brief periods of increase in good years alternate with periods of decline in poor years. On the whole density is held well below carrying capacity but occasionally particularly favourable periods may lead to outbreaks or unfavourable periods lead to local extinction. Density-dependent effects may become important at such times.

At a particular level these two rates would be equal and a stable population would be reached. This level has sometimes been called the *carrying capacity* (K) implying that the environment is full. The carrying capacity is often presented as a constant in theoretical treatments, which may be true for resources such as nest-holes in trees but is clearly not for food. Thus one can envisage a population showing widely fluctuating birth and death rates and pursuing an elusive stable level (Figure 32b).

Andrewartha and Birch stressed that in most animals for most of the time conditions were unfavourable for breeding, so that populations would be declining slowly. Occasionally the weather, or some other factor may cause high mortality whereas on other occasions conditions may be good enough to allow breeding. Populations then rise, sometimes rapidly, but only until conditions again become unfavourable. They believed that only rarely would resources become short due to the population itself. Normally climate, acting in a manner independent of density, would hold populations well below the carrying capacity (Figure 32c).

Lack (1954, 1966) extended these more general theories of population regulation to birds and strongly supported Nicholson's point of view. He maintained that the reproductive rates of birds are at the maximum that the environment and the abilities of the birds allow. Mortality rates on the other hand tend to balance birth rates in a density-dependent manner so that populations tend to remain stable. He proposed that food was the ultimate limiting factor in most cases, though predation or disease could hold populations below the level at which food became scarce. I have already discussed in Chapter 5 the evidence that birds lay clutches that will give them the greatest number of offspring. With a few modifications this hypothesis was supported. The question of mortality rates being related to density was on the whole supported by the studies presented in Lack's (1966) book. I shall present in some detail the argument for European titmice as recently summarized by Perrins (1979) (see also Perrins and Birkhead 1983).

Great Tits (*Parus major*) near Oxford show fluctuations in density over a period of years. In years of high density they lay smaller clutches, have fewer second clutches, hatch a smaller proportion of eggs and therefore rear fewer young per pair. Higher densities tend to encourage predators such as Weasels (*Mustela nivalis*). There is an indication that recently fledged young survive less well at high densities. The autumn appears to be a difficult time for young titmice, and at least in the Netherlands, survival is better when the population is lower. Kluijver (1966, quoted by Perrins 1979) found that Great Tits on a Dutch island fluctuated in a rather regular fashion for several years. For five years he removed 60% of nestlings and found that the population still fluctuated around the same level. There is thus considerable evidence of density dependent factors operating on tit populations. However, there are also density independent factors as well, including winter food supply. Great Tits, in particular, feed on the nuts (mast) of the beech tree (*Fagus sylvaticus*) and these are produced abundantly only in some years. In about half of the years virtually no nuts are produced. Great Tit populations tend to decline after such nut failures and increase with a bumper crop of nuts. Survival during winter can also be affected by severe winter weather in Britain, though on the whole mean winter temperature and breeding density in the following season are not correlated. However, there is such a correlation in Finland, which experiences colder winters than Britain (von Haartman 1971). In many areas there are no beeches, in particular the northern parts of the Great Tit's range, and there populations fluctuate in a similar way to those in areas with beeches. Again autumn food supply may be implicated.

Von Haartman (1971) stressed that resources other than food can be important in regulating breeding populations. The most obvious of these is nest-holes for hole-nesting birds, but other species require dense shrubs to breed safely and a few species require features such as song-posts in open habitats. Most studies on titmice have been on populations nesting in boxes. Another popular species for study in northern Europe, the Pied Flycatcher (*Ficedula hypoleuca*), is also a hole-nester. Holes in trees are generally uncommon in managed or secondary forest, virtually the only kind in Europe and provision of nest-boxes will almost invariably increase the abundance of birds such as tits and Pied Flycatchers. It is probable that holes are more common in natural forests, as there are more very old or dead trees but it is not known whether they are sufficient even there. Certainly tits and other hole-nesters are a rather minor element of the avifauna of the primeval Bialowieza Forest in Poland (Tomialojc *et al.* 1977). Nest boxes are usually safer from predators than are natural holes and they are less likely to become water-logged. Indeed newer designs are being developed to deter weasels and woodpeckers. Perhaps, then, it is not surprising that food and winter weather emerge as the main factors that limit these populations. Even fluctuations in these are dampened by man's activities as now many titmice visit suburban bird tables for nuts and fat, especially in hard weather.

Another important concept in population ecology is that of *self-regulation*. Do populations tend to control themselves below the level at which food shortages occur? Wynne-Edwards (1962, 1986) is the champion of self-regulation in populations of birds and other vertebrates. He argued that a suite of behaviours, including territoriality, has evolved in order to hold down birth rates at approximately the level of death rates. This would require some individuals to delay or forgo reproduction or at least to produce clutches smaller than their capacity. In addition, individuals tend to leave areas of high density, often *en masse*, thereby avoiding over-exploitation of food. Such phenomena appear to require the operation of group selection which, as argued earlier, is unlikely to be stronger than individual selection. Wynne-Edwards (1986) pointed out that most vertebrate populations consist of small groups between which there is little gene-flow. Those that conserve their resources best should last longer. One would still expect those individuals reproducing at a faster rate to swamp the more conservative individuals.

However, there is a certain amount of truth in what Wynne-Edwards said. Studies carried out over more than 20 years on Red Grouse (*Lagopus scoticus*) have shown that territorial behaviour spaces out breeding pairs causing excess birds to emigrate to areas where prospects of survival are poor (summarized by Wynne-Edwards 1986). The basis of this, though, appears to be natural selection operating directly on individuals. Those that obtain territories breed; those that try but fail refrain from further, fruitless competition. They simply try to survive in suboptimal areas until the next breeding season.

It appears that populations tend to be stabilized by territorial behaviour, but this may be a by-product of such behaviour rather than the reason it evolved in the first place.

Lack, especially in his 1966 book, argued convincingly that resources, such as food, may ultimately limit bird populations. In most cases predators and disease tend to take excess birds. Weather may at times devastate populations but they normally recover within a year or two. Constraints imposed by the environment tend to adjust mortality rates to balance birth rates on average. Birth rates, though, are modified less by the environment.

One question that is less often asked is whether indeed populations are stable. In fact what is meant by stability? Birds of prey and herons may maintain their populations at remarkably constant levels over many years (Newton 1979; Lack 1966). Moreover, they show an amazing ability to recover from catastrophes (i.e., resilience — see Chapter 4). Grey Herons (*Ardea cinerea*) are severely affected by cold winters which may remove two-thirds of the breeding birds, but they recover rapidly (Figure 33). The Peregrine Falcon (*Falco peregrinus*) was decimated in

Fig. 33. Numbers of breeding pairs of Grey Heron in Britain (from Sharrock 1976).

southern England during the Second World War because it captured message-carrying pigeons. It was well on the way to recovery by 1950 when it was almost exterminated by the build up of chlorinated hydrocarbons. With the banning of these pesticides it has now returned to pre-war levels in most areas. Self-regulation may play an important role in such large predatory species.

Smaller birds on the whole show larger fluctuations that are less easily explained. Local populations rise and fall and occasionally go extinct. On a larger scale the British Trust for Ornithology has been monitoring bird populations for over 20 years (O'Connor 1985). Each year about 60 species are given an index based on the 1966 figure, which is taken as 100. In 1985 no less than 13 species were at levels greater than double their density in 1966 and eight were below half their 1966 density (Marchant and Whittington 1986). Most of these changes represent progressive rises or declines rather than just the peaks and troughs of fluctuations. The increasing species include the Collared Dove (*Streptopelia decaocto*), which introduced itself in the 1950s and is now one of the commonest birds of the British countryside (Figure 34). On the debit side many migrants that winter in sub-Saharan Africa have declined dramatically. A few rare species like the Red-backed Shrike (*Lanius collurio*) have steadily declined this century to virtual extinction in England.

Extensive censuses have also been carried out periodically in the coniferous forests of Finland (Väisänen *et al.* 1986). Progressive declines and increases have also been found there in many species. In fact only eleven species were classed as "stable or fluctuating", whereas sixteen had decreased and another sixteen had increased over three decades. One of the increasing species was the Wryneck (*Jynx torquila*), which has virtually disappeared from England. Winter cenuses in Finland also revealed many increases and decreases (Tiainen 1985). Enemar *et al.* (1984) found similar patterns in birch forests in Swedish Lapland over 20 years. Although a few species responded to local changes in the abundance of food such as caterpillars and birch seeds, the majority fluctuated in ways not obviously related to local conditions. The authors concluded that conditions during migration or on the wintering grounds had a greater influence on population sizes, especially in long distance migrants, which showed an overall decline.

It could be argued that most of these changes are the direct or indirect result of human activity, such as changes in land-use, and overgrazing in the Sahel region of Africa. Also, many species are at the edge of their distribution in these countries so that a slight deterioration or amelioration of climate will affect their densities disproportionately there, compared with the centre of their ranges.

Some species appear to make a habit of erratic population fluctuations, often accompanied by unpredictable movements, which resemble the desert locusts studied by Andrewartha and Birch (1954). The Boreal Forests contain many such species, including Crossbills (*Loxia*), Redpolls and Siskins (*Carduelis*), Nutcrackers (*Nucifraga*) and Waxwings (*Bombycilla*) (Larson and Bock 1986, and papers cited by van Haartman 1971). All of these species are seed or fruit-eaters whose food supply fluctuates greatly and often synchronously. Failure of the seed or fruit crop leads to population crashes and mass emigration. Plentiful food leads to high survival, successful reproduction and residency. As the birds cannot affect how much seed will be produced each year, this is clearly a density-independent factor. Density-dependent factors may operate, for instance in the breeding season, though

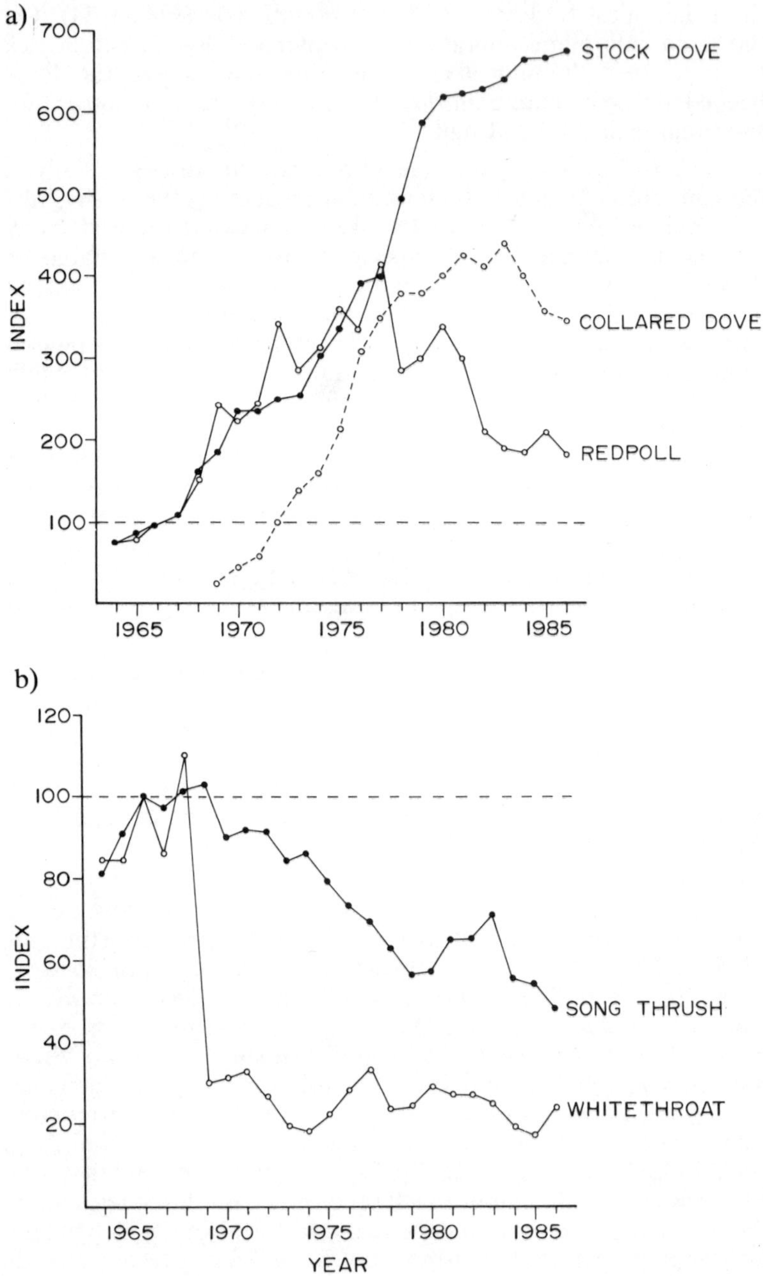
a)
700
600
500
400
300
200
100
INDEX
1965
1970
1975
1980
1985
STOCK DOVE
COLLARED DOVE
REDPOLL
b)
120
100
80
60
40
20
INDEX
1965
1970
1975
1980
1985
YEAR
SONG THRUSH
WHITETHROAT

these irruptive species are not notably territorial and perhaps breed only when food is abundant. Possibly the few remaining individuals in years of mass emigration are able to exploit or defend the remaining food sources. Certainly this group of species merits closer study, but perhaps their population dynamics approach the pattern shown in Figure 32c, though a widely fluctuating carrying capacity is more likely (Figure 32b).

The influence of genetic factors on population changes should not be ignored. Perhaps species that suddenly expand their ranges have produced a new phenotype that allows them to cope with different food or withstand greater physiological stress or changes their behaviour to tolerate human presence. Several species have altered their winter diet and even their migratory patterns in response to food provided by people in Europe. It is not known to what extent this is due to learnt or inherited changes and how much the behavioural changes have been accompanied by physiological changes. Survival over the winter will have increased, probably leading to greater breeding populations or limitation at another time of year.

Population Ecology of Australasian Birds

Seabirds

Because seabirds have low adult mortality rates, modest reproductive rates and deferred maturity their populations do not show the level of fluctuations found in landbirds. This is not to say that their populations are stable. Gannets (*Morus bassana* and *M. serrator*) in both the North Atlantic and New Zealand are steadily increasing (Nelson 1964; Waghorn 1983), as are many other seabirds. This is at least partly because their colonies now receive less human persecution and disturbance, but also because their

---

Fig. 34. opposite. Examples of species that have changed greatly in density in the United Kingdom in recent decades (from successive issues of *Bird Study* and *BTO News*, but see also Sharrock 1976).

a) Two doves (*Columba oenas* and *Streptopelia decaocto*) that rose steadily for over ten years before possibly reaching a new stable level, and the Redpoll (*Acanthis flammea*) which rose for ten years and is now falling again (see O'Connor and Shrubb 1986 for effect of farming changes on the doves).

b) The Whitethroat (*Sylvia communis*) a trans-Saharan migrant whose population crashed in 1969 and has not recovered, deterioration of winter habitat and possibly loss of hedges in breeding areas may be the cause, and the Song Thrush (*Turdus philomelos*) which has declined steadily in woodland possibly due to a decline of spring sowing of cereals (O'Connor and Shrubb 1986).

favoured food may be increasing as a result of other competing species being overfished by humans. Periodically seabirds suffer crashes in their populations resulting from severe weather (Underwood and Stowe 1984), diseases (e.g., Sutcliffe 1986), outbreaks of poisonous shellfish (Coulson *et al.* 1968) or food shortages, for instance caused by warm water in El Niño years (Hughes 1985). Locally, seabird populations can increase rapidly though immigration, e.g., Puffins (*Fratercula arctica* – Harris 1983).

Several species of seabird breeding in Australia and New Zealand have received close and long-term study. Richdale's (1957) work on the Yellow-eyed Penguin (*Megadyptes antipodes*) near Dunedin was a pioneering study in population ecology (Table 36, Plate 40). About one quarter of the young birds reared in the colony were recovered breeding, which would give a minimum recruitment of about 9%. As some birds would have moved to other colonies and others immigrated this would mean that recruitment rate and adult mortality rate would be rather closely balanced. However, through his 18-year study Richdale's colony initially declined then gradually increased to about three times its initial size. There seemed to be little evidence that the rate of increase was slowing as density rose.

The Little or Blue Penguin (*Eudyptula minor*) has also been studied in Australia and New Zealand. The colony at Phillip Island in Victoria is a tourist attraction and young and adults have been banded there over a long period (Reilly and Cullen 1979, 1981, 1982). There is no evidence that the population size is changing markedly despite problems with dogs, traffic and occasional wilful killing by people. Adult death rate is similar to that in the Yellow-eyed Penguin though productivity of young is almost double that of the latter. Only a very small proportion of chicks have been recovered in later years at the breeding colony, but this is clearly an underestimate as only a section of a large colony was studied in detail. A colony in Tasmania, however, was declining in the 1960s as mortality rates of adults was about three times as great as on Phillip Island whereas production of young was lower (Hodgson 1975 in Reilly and Cullen 1979). There was an indication that some birds were starting to breed at a younger age than they do on Phillip Island. Delayed maturity and breeding can have a major impact on the potential rate of increase of a population. One way in which seabird populations can be regulated is by altering the age of first breeding. This can be lowered when populations are low or declining but increased at high densities. Large scale mortality of Little Penguins can also occur in the

Table 36. Some population data on a range of Australasian seabirds.

	Adult mortality rate	Young reared/ pair/year	Yearling survival	Age at first breeding	Banded young recaptured at breeding colony	Population trends
Yellow-eyed Penguin (Richdale 1957)	14%	0.36	?50%	2–3 yrs 3–4 yrs	26%	Fall then steady rise
Little Penguin:						
Victoria – (Reilly and Cullen 1979, 1981, 1982)	14%	0.71	?80%	3 yrs	2.8%	Stable?
Tasmania – (Hodgson 1975)	42%	c.0.40	?	<3 yrs	?	Declining
Short-tailed Shearwater (Serventy 1967)	7%	?	>50%	5–8 yrs	44%	Stable?
Australasian Gannet (Nelson 1978; Waghorn 1983; Wingham 1984)	4–5%	0.61	?	4–7 yrs	?	Steady to rapid increase

non-breeding season, possibly through starvation and disease (Delroy 1985).

Short-tailed Shearwaters *(Puffinus tenuirostris)* and indeed many other seabirds show very low adult mortality rates as well as delaying breeding up to the age of ten years (Serventy 1967; Perrins and Birkhead 1983). As they lay a single egg, which is not replaced if lost, their productivity is not high. Mortality of the deserted young can be high soon after they leave the colony so that ultimate recruitment will also be low. Despite harvesting of the young Short-tailed Shearwaters (muttonbirds) in the tens of thousands each year there is no indication that populations are declining. Again adjustment of the age at first breeding is the most likely way in which recruitment could match mortality.

Some seabirds have declined, but usually due to some human agency. Rats have been introduced onto islands off Australia and New Zealand and have the ability to decimate breeding colonies. Abbott's Booby is threatened on Christmas Island in the Indian Ocean by clearing of forest trees, in which it breeds, to mine phosphate (Nelson and Powell 1986). Phosphate mining ceased in December 1987, though breeding near mined areas is still depressed (Stokes 1988). Populations on other tropical islands are directly depredated by man.

In contrast, other seabirds, particularly gulls, have increased dramatically by turning from food gathered at sea to human refuse. The Silver Gull (*Larus novaehollandiae*) is displaying a population explosion in parts of Australia similar to that shown by some European and North American gulls (Gibson 1979). Such increases may be to the detriment of other seabirds such as terns and petrels.

Waterbirds

Water is rarely in plentiful supply in most of Australia; consequently many waterbirds have to breed in an opportunistic fashion. The Grey Teal (*Anas gibberifrons*) has been well studied (Frith 1967). It moves about nomadically towards, or in search of, areas where water levels are rising (see Chapter 9). One of the most reliable areas is the Murray Darling basin of inland New South Wales, southern Queensland and northern Victoria. Here, rivers flood at least every two or three years resulting from rain or melting snow from the Great Dividing Range. Mature billabongs and swamps, in which the water levels are rising, attract

Grey Teal, which immediately start to breed. Aquatic insects and other invertebrates become abundant and are eaten by the ducklings whereas the adults, and later the young birds, consume more aquatic plants. As water levels decline and pools dry up, the food supply declines and the Teal eventually leave in search of more productive areas. Grey Teal can breed annually in well watered, particularly coastal, areas. In favourable years populations expand rapidly, and exceptionally breeding takes place almost throughout inland Australia. However, in dry years breeding may be minimal and when two or more years of drought occur sequentially mortality increases and the populations decline. In long droughts they may take refuge in saltwater habitats where they develop a nasal gland that excretes excess salt (Lavery 1970). Grey Teal, and some other waterfowl, are largely at the mercy of extrinsic factors, principally rainfall. Their population ecology appears most like Figure 32c, a pattern that usually is more appropriate for insects than for birds.

Australian Magpie

One of the longer population studies, which has not yet been fully published, is that of Robert Carrick on the Australian Magpie (*Gymnorhina tibicen*) (Carrick 1972). Magpies are common in open habitats throughout much of Australia providing there are some trees. Carrick's study area near Canberra was on farming land with scattered trees on the better land and woodland on the hills. Magpies belong to several types of groups (Table 37). The

Table 37. Frequency of different groups found in Australian Magpies (from Carrick 1972).

	Permanent	Marginal	Mobile	Open
Group composition				
1♂ 1♀	47%	53%	20%	36%
1♂ 2♀	35%	33%	30%	31%
2♂ 1♀	5%	1%	25%	22%
>2♂ or >2♀	13%	14%	25%	11%
Habitat: Adequate tree cover	100%	39%	80%	0%
Mean duration of tenure	>47 months	11 months	8 months	8 months
Productivity (Independent young/adult)	0.32	0.09	0	0

highest ranking birds live in groups that defend permanent territories in which they breed. Defence is vigorous and intruders are driven out and in the breeding season predators, including humans, are attacked. About half of the pairs are attended by additional birds, often females, though usually only one female breeds successfully. The best territories include open grassland for foraging and tall trees for nesting. Poorer sites with fewer trees or smaller foraging areas are occupied by marginal groups, which may attempt to breed but are rarely successful. Such territories do not last as long as those of permanent groups. Additional birds belong to mobile groups, which move around among poor quality areas and do not attempt to breed.

Finally, the remainder belong to the flock, whose size varies and which occupies open treeless areas. Members of the open flocks do not breed and are typically young birds that have been evicted from the parental territory.

The number of permanent groups was nearly constant throughout the twelve years of the study, ranging from 34 to 39. Indeed eight of the initial 37 groups were present at the end of the study. These groups had 37–54 adult males and 48–66 adult females. Permanent groups produce most of the offspring in the population, which mostly spend about three years in the non-territorial flock before gaining territorial status. Survival is high in the permanent groups so that vacancies are few (10 of the original 104 birds survived the whole study). Sometimes such groups break up and their members revert to the flock before steadily working their way back up the hierarchy.

The reproductive rate of the population as a whole is low, partly because there are birds that do not occupy breeding territories, but also because some individuals in good territories do not breed. Many clutches are lost to ravens or bad weather. Recently fledged young fall prey to cats and foxes and many juveniles are killed by vehicles.

Survival of flock members is much lower than that of territory holders with many deaths being attributed to humans or other predators. In one wet winter many flock birds died of disease, which did not affect birds occupying territories. Some flock members are dangerously light in weight by late winter, though territorial birds rarely reach this level.

Overall, the breeding population of Australian Magpies is maintained by strong territorial behaviour in the best habitats for breeding and survival. In this way a surplus of non-breeding marginal or non-territorial birds is maintained. This surplus fluctuates more than the breeding population and there is an indication that mortality within this segment is to some extent density-dependent. Disease may be more likely to break out at high density and competition for food may lead to weight loss and death.

Heron Island Silvereyes

Jiro Kikkawa and his students have studied Silvereyes (*Zosterops lateralis*) on Heron Island for over two decades and have been able to demonstrate how their population is regulated (Kikkawa 1977, 1980; Kikkawa and Wilson 1983; Catterall *et al.* 1982; Catterall 1985). There are major advantages in studying populations on islands. Their geographic limits are predetermined, and generally emigration and immigration are minimal. There is a real chance that one can get to know every bird. The factors limiting island populations may be different from those limiting mainland populations as predators may be absent or scarce, as will be competitors. Populations on islands thus become generalized in foraging behaviour and can reach high densities. This predisposes them to food shortages or the development of self-regulatory behaviour.

Silvereyes are interesting in that they are very generalized foragers that have colonized Australia from Asia over the last few million years and New Zealand over the last century and are remarkably successful in a wide range of habitats.

The numbers of Silvereyes on Heron Island fluctuated by a factor of two over nine years (Figure 35). Low breeding populations followed two years in which there were cyclones (January 1967 and April 1972). These caused high mortality of recently fledged young in the first case and birds of all ages in the second case. The population recovered within a year after the cyclones. This was mainly because there was lower mortality in the winter following the cyclone and a higher level of recruitment after the next breeding season. There was considerable evidence that food, chiefly figs and insects, were limiting in winter and the breeding season. Silvereyes were territorial with dominant birds holding territories in areas near to the best supply of the energy-rich figs. Dominant birds tended to survive winter better than more subordinate birds (Table 38). Subordinates were less likely to breed, but the occupation of good territories did not ensure that

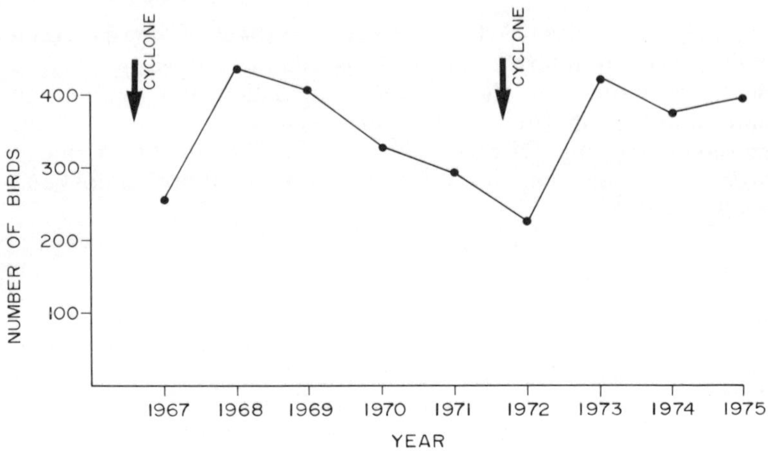

Fig. 35. Numbers of Silvereyes on Heron Island at the beginning of the breeding season over nine years (from Kikkawa 1977).

dominant birds had the best breeding success (Table 39). This was because territories were more tightly packed when good sources of food were available nearby.

Heron Island Silvereyes are larger than their mainland counterparts, probably because their high density imposes strong selection in favour of dominance. The density of this single species (about 25 birds/ha) exceeds that of the whole bird assemblage in many mainland habitats.

To summarize: the weather can impose high mortality and reduced breeding density on the Heron Island Silvereye population. Density-dependent mortality based on winter food abundance and to some extent density-dependent breeding productivity due to territoriality and competition both tend to stabilize the population within a rather narrow band.

Table 38. Survival during winter (May-August 1968) of Silvereyes of different status on Heron Island (from Kikkawa 1980).

	Dominant	Intermediate	Subordinate
% surviving: First-year birds	91%	83%	78%
Adults	94%	90%	89%
% gaining 0.5 g+	31%	28%	24%
% losing 0.5 g+	11%	17%	32%

Table 39. Breeding success of Heron Island Silvereyes of different status (from Kikkawa and Wilson 1983).

	Dominant	Intermediate	Subordinate
% known to have bred	79%	61%	52%
% of young surviving to end of April 1970	46%	62%	41%

Other landbirds

Several other species of landbirds in Australia have been the subject of population studies, though many have been of rather short duration. Fairy-wrens were discussed extensively in the previous chapter. The number of breeding groups does not change greatly over a period of years, principally because the best habitat is defended. Suitable habitat is fully occupied in most years, though there is some expansion and contraction of territory size. The total adult population fluctuates rather more because the number of non-breeding helpers per group is variable. Thus self-regulation tends to ensure that mortality from and recruitment into the breeding population is balanced. A similar mechanism probably operates in many co-operative breeders.

Many small birds of eucalypt forests and woodlands display low adult annual mortality rates (20–45%) which, along with rather low breeding productivity (1–3 fledglings per season), leads to dampened fluctuations between years (Table 40). Most of these species are territorial and even in those that are not co-operative breeders there is probably a surplus of non-breeding individuals. On occasion, though, populations may decline markedly. During the early 1980s a severe drought struck much of eastern Australia. Some species decreased during this period, whereas others were not seriously affected (Table 41). Bell (1985b) studied three species of thornbills through this period and his results indicated that the drought affected populations more by reducing breeding success than by increasing mortality. The number of breeding attempts was reduced and few young were reared to independence. Other species respond to drought by emigrating.

Populations of nectar-feeding birds also include large numbers of non-territorial individuals. Paton (1985a) found that New Holland Honeyeaters (*Phylidonyris novaehollandiae*) defended territories for most of the year in Victoria, but there were many transients, particularly in late spring and early summer, after the breeding season. Many of these were juveniles, which

Table 40. Adult mortality rates and breeding productivity (expressed various ways) in some Australian passerines.

Species	Adult mortality rate	Breeding productivity
Eastern Yellow Robin (marchant 1985)	25%	0.74/nest
Treecreepers (3 spp.) (Noske 1983, 1985)	20–30%	1.0–1.4/group/season
Brown Thornbill	13%	0.56/nest
Buff-rumped Thornbill (Bell 1985b)	46%	1.20/group/season
Superb Fairy-wren		
(Rowley 1965)	34%	1.70/nest
(Nias 1987)	55%	2.70/group/season
Splendid Fairy-wren (Rowley 1981)	43%	3.24/group/season
Brown Honeyeater (*Lichmera indistincta*) (Robertson and Woodall 1987)	44%	Not calculated
New Holland Honeyeater (Paton 1985a,b)	45%	3/pair/season
Australian Magpie (Carrick 1972)	15%	0.32/adult/season (permanent groups only)

Table 41. Changes in estimated densities (birds/ha) of chosen species in eucalypt woodland during a severe drought from 1979 to 1981 (Ford *et al.* 1985).

	Species that apparently declined		
	1978	1979	1981
Golden Whistler	0.43	0.79	0.04
Buff-rumped Thornbill	1.54	0.92	0.21
Striated Thornbill	1.58	0.25	0.06
Brown-headed Honeyeater	0.65	0.38	0.04
Silvereye	0.36	0.88	0.01
Diamond Firetail	1.18	0.25	0.05
	Species apparently not affected		
	1978	1979	1981
Eastern Yellow Robin	0.68	0.67	0.39
Superb Fairy-wren	1.03	1.41	1.18
Speckled Warbler	0.49	0.54	0.41
Varied Sittella	0.20	0.20	0.30
Fuscous Honeyeater	3.07	5.46	3.23

showed lower survival than did adult birds. Perhaps only a few of these surplus birds survived to the next breeding season.

John Le Gay Brereton studied Eastern Rosellas (*Platycercus eximius*) near Armidale over many years. Details of this intensive population study have unfortunately never been published, but the main findings have been summarized (Brereton 1963, 1971; Smith and Brereton 1976; Wyndham and Cannon 1985). Breeding adults are sedentary and long-lived. Juveniles disperse in late summer and join flocks, which can be large. Although Rosellas are capable of breeding in their first year most young birds remain in non-breeding groups for at least their first spring.

The population typically consists of a rather stable territorial segment and a more variable non-breeding component. A similar pattern may be found in other parrots living in woodlands or savanna. The Budgerigar (*Melopsittacus undulatus*), which inhabits semi-arid habitats, displays a different population structure. It is not territorial but semi-colonial when breeding, and most or all individuals apparently attempt to breed (Wyndham 1982). Budgerigars display very marked local fluctuations in density. These are at least partly due to migration and nomadism, but they also suffer high mortality during droughts. Their more unpredictable environment appears to have led to a reduced dependence on self-regulation and perhaps an increased density independent mortality.

Summary

Seabirds, in Australia and elsewhere, tend to have low birth and death rates and consequently their populations do not vary much between years. Despite this, many populations appear to show a pattern of steady gradual increase interspersed with brief periods of often dramatic decline. Although density-dependent factors undoubtedly operate, this pattern bears some resemblance to Andrewartha and Birch's model of steady decline with brief periods of rapid increase.

Australian waterbirds seem to be dependent on external conditions, principally rainfall leading to high water levels and abundant food, with perhaps only rare shortages induced by the populations themselves.

Overall, the studies to date indicate that many Australian landbirds maintain populations that do not fluctuate widely. The breeding populations, in particular, are often remarkably stable.

This is largely the result of self-regulation, strong defence of prime breeding habitat and the presence of a non-breeding surplus. These extra birds may live in breeding territories as non-breeding helpers or may occur in territories in marginal habitat or in roving flocks. There is some evidence that food is a limiting factor both in holding down reproductive rates (see Chapter 5) and in affecting winter mortality. With rather low reproductive and adult mortality rates perhaps few birds actually starve. Predation, weather and disease take a steady toll, particularly of non-breeding birds. I do not wish to imply that population control is qualitatively different in Australian birds from birds of northern temperate regions. It is more that the emphasis tends to be on an adjustment of per capita birth rate through behavioural means, rather than on high density-dependent death rates. European birds are often territorial and excess non-breeding individuals are frequently present, but over-winter mortality resulting from food shortage probably has a greater influence on the size of the breeding population than in Australia.

However, one should be cautious of drawing premature conclusions from a small number of perhaps atypical species. Most birds that have been studied, in any continent, tend to be sedentary or at least regular summer visitors, which are reasonably common and predictable. Few post-graduate students would take on a species that is rare at the best of times and may even be absent next year. Just as Budgerigars may well be regulated in a different way from other parrots, if indeed their populations can be said to be regulated, so may many passerines differ from fairy-wrens, magpies and silvereyes. I should like to see studies of species like chats, songlarks and woodswallows where there can be great fluctuations at least locally, also of species like Regent Honeyeaters (*Xanthomyza phrygia*), which appear to be common nowhere. How much do species like Gibberbirds (*Ashbyia lovensis*), grasswrens and wedgebills, which inhabit remote, arid areas, fluctuate in abundance and what determines their density? Who is game to tackle species like the lorikeets, cuckoos, emu-wrens and quail-thrushes?

1

2

3

4

5

6

Plate 1: A sandy desert in northwestern New South Wales (Photograph by H. Heatwole); **Plate 2:** Striated Grasswren, found in arid, sandy areas where Spinifex (*Triodia*) dominates. (Photograph by W. Labbett); **Plate 3:** Gibber plain, interspersed with sand-hills (eastern Simpson Desert, Queensland), note standing water after unusual rain (Photograph by H. Heatwole); **Plate 4:** Succulent shrub-steppe with Blue-bush (*Maireana*), (Gawler Ranges, South Australia). (Photograph by H. Ford); **Plate 5:** Crimson Chat, a honeyeater commonly found in arid shrublands. (Photograph by G. Little); **Plate 6:** Acacia woodland (*A. sowdenii*) (Eyre Peninsula, South Australia). (Photograph by H. Ford).

7

8

9

10

11

12

Plate 7: Arid acacia and eucalypt woodland (Macdonnell Ranges, Northern Territory). (Photograph by H. Ford); **Plate 8:** Mallee, multi-stemmed eucalypt woodland, with grassy understorey (near Blanchetown, South Australia). (Photograph by H. Ford); **Plate 9:** Malleefowl, a megapode that inhabits mallee and other dry eucalypt woodland. (Photograph by R. Miller); **Plate 10:** Eucalypt woodland (Northern Tablelands, New South Wales). (Photograph by H. Ford); **Plate 11:** Woodland suffering from eucalypt dieback (Northern Tablelands, New South Wales). (Photograph by H. Ford); **Plate 12:** Laughing Kookaburra, common in woodland and forest. (Photograph by H. Ford).

13

14

15

16

17

18

Plate 13: Open forest (wet sclerophyll) (Canberra, Australian Capital Territory). (Photograph by H. Ford); **Plate 14**: Open forest of Brushbox (*Lophostemon*) (Lamington National Park, Queensland). (Photograph by H. Ford); **Plate 15**: Crimson Rosella, in banksia, but typically a species of forests. (Photograph by D. Turner); **Plate 16**: Ecotone between banksia heathland and cool temperate rainforest (*Nothofagus*) (New England National Park, New South Wales). (Photograph by H. Ford); **Plate 17**: Male Superb Lyrebird displaying. (Photograph by L. Schick); **Plate 18**: Sub-tropical rainforest (Dorrigo, New South Wales). (Photograph by H. Ford).

19

20

21

22

23

24

Plate 19: Noisy Pitta, member of a group found throughout the Old World tropics. (Photograph by R. Hardie); **Plate 20:** Banksia heathland on Fraser Island (Photograph by H. Heatwole); **Plate 21:** White-cheeked Honeyeater, common in coastal heathland north of Sydney and in southwestern Australia. (Photograph by R. Howe); **Plate 22:** Mangrove on Saibai, Torres Straits (Photograph by H. Heatwole); **Plate 23:** Striated (Mangrove) Heron. (Photograph by I. Frien); **Plate 24:** Part of the extensive seasonal wetlands in the East Alligator River catchment, Northern Territory. (Photograph by R. Simpson).

25

26

27

28

29

30

Plate 25: Comb-crested Jacana. (Photograph by R. Hardie); **Plate 26:** Royal Spoonbill. (Photograph by H. Ford); **Plate 27:** Palaearctic waders at Roebuck Bay, Western Australia. (Photograph by C. Minton); **Plate 28:** Sooty Oystercatcher, endemic wader of rocky coasts. (Photograph R. Hardie); **Plate 29:** Cape Barren Goose, which breeds on small islands off southern Australia. (Photograph by H. Ford); **Plate 30:** Masked Boobies on Gannet Cay, Great Barrier Reef. (Photograph by H. Heatwole).

31

32

33

34

35

36

Plate 31: *Astroloma conostephioides,* a tubular flower visited mostly by long-beaked honeyeaters. (Photograph by H. Ford); **Plate 32:** *Eucalyptus diversifolia,* nectar in eucalypt flowers is easily accessible to a wide range of birds, and other animals. (Photograph by H. Ford); **Plate 33:** *Banksia spinulosa,* brush inflorescence, which can produce large quantities of nectar. (Photograph by H. Ford); **Plate 34:** Male Rufous Whistler. (Photograph by G. Weber); **Plate 35:** Male Satin Bowerbird, note female in bower. (Photograph by C. Mulquiney); **Plate 36:** Australian Brush-Turkey. (Photograph by R. Hardie).

37

38

39

40

41

42

Plate 37: Superb Fairy-wren. (Photograph by C. Mulquiney); **Plate 38:** Eastern Yellow Robin (note metal and colour bands). (Photograph by C. Mulquiney); **Plate 39:** Group of Grey-crowned Babblers, adults (3 years plus) have yellow irises, younger birds have brown irises. (Photograph by G. Chapman); **Plate 40:** Yellow-eyed Penguin. (Photograph from Otago Peninsula Trust); **Plate 41:** Pied Currawong, an altitudinal migrant. (Photograph by H. Ford); **Plate 42:** Silvereye, a partial migrant. (Photograph by C. Mulquiney).

43

44

45

46

47

48

Plate 43: Yellow-faced Honeyeater, a regular migrant in southeastern Australia. (Photograph by C. Mulquiney); **Plate 44:** White-throated Gerygone, a north–south migrant within Australia. (Photograph by H. Ford); **Plate 45:** Sacred Kingfisher, in which many breeding birds in southern Australia winter in New Guinea and Indonesia. (Photograph by R. Hardie); **Plate 46:** Wedge-tailed Shearwater, Australian breeding birds winter off the Philippines. (Photograph by H. Ford); **Plate 47:** White-browed Woodswallow, a regular north–south migrant which often irrupts beyond its normal range. (Photograph by C. Mulquiney); **Plate 48:** Grey Teal, a nomadic species that invades central Australia after heavy rain, but retreats to the coast during drought. (Photograph by H. Ford).

9. Migration

Types of Movement

The journeys of birds take a wide variety of forms. Even many of the most sedentary species will wander out of their breeding territories during autumn and winter. Often they join up with members of their own species or different species to form roaming flocks. Typically these species return to their territories to roost and may sing or display there in the early morning or evening.

Young birds usually disperse from their parental territory, travelling hundreds of metres to a few kilometres in search of suitable habitat. Adults too may disperse quite widely after breeding, especially if breeding requires particular features such as nest-holes or dense shrubs. Many forest species occupy more open habitat during the non-breeding season. They may cover a large area in search of concentrations of food. Such dispersal is also shown by many seabirds, which aggregate in a few large colonies to breed but spread out into the surrounding ocean during the rest of the year. Many seabirds display true migration though.

Migration can be defined as the regular and repeatable movements of populations between different geographical areas. These can be as short as a few kilometres in the case of some altitudinal migrants, which move from mountains to lowlands or the coast. Several waders like Northern Pied Oystercatchers (*Haematopus ostralegus*) and Western Curlews (*Numenius arquata*) do this in western Europe. Alternatively, they may be thousands of kilometres as in the Arctic Tern (*Sterna paradisaea*) which nests in the arctic but spends the non-breeding season in the Southern Ocean. Adults may perform a round trip of 35,000 kilometres during which they experience no winter. Many migrants are faithful to their breeding or natal area, returning to the same territory or even nest in successive years. Sometimes they may use the same wintering area repeatedly. More often though they wander during the non-breeding season tied to a nest or young, and their food may be more patchily distributed.

Sometimes long distance movements are not regular and the presence of a species in an area is unpredictable. In the last chapter I mentioned several species of northern seed-eaters and frugivores which often irrupt south beyond their usual wintering grounds. Siskins (*Carduelis*), crossbills (*Loxia*) and waxwings (*Bombycilla*) migrate only moderate distances in some years but, when the seeds and berries that they eat have failed, may travel hundreds of kilometres. These movements are predictable in that they occur in early winter often after typical migrants have already left and the survivors almost invariably return to the same breeding area. There are exceptions; for instance, newly established pine plantations in England now have Common Crossbill (*Loxia curvirostra*) populations, which originated from winter irruptions (Newton 1972). Some birds of prey may also irrupt, especially species that are dependent on small mammals like lemmings and voles whose numbers vary dramatically. Rough-legged Buzzards (*Buteo lagopus*) and Snowy Owls (*Nyctea scandiaca*) are predators of rodents that irrupt in North America and Eurasia.

Somewhat different from irruption is *nomadism*. At its most extreme nomadism can occur at any time of year and involves the movement of a significant proportion of the population outside its normal range. Such birds may actually breed in the new areas and remain for several years before dying out or departing and are not seen again for years or even decades. Nomadic species typically inhabit grasslands and deserts where a succession of good years may lead to population increases. A deterioration in conditions and ensuing food shortage may then result in extensive departures. The classic nomad to Europe is Pallas's Sand-grouse (*Syrrhaptes paradoxus*), which inhabits the semi-arid steppes of central Asia. Periodically it sweeps westwards and has been known to breed in Britain.

In Africa some species of queleas (*Quelea*) are described as nomads, breeding rather unpredictably in most areas. However, some populations at least show itinerant breeding, regularly nesting in one area after rain then departing *en masse* to re-nest in a second region (Jaeger *et al*. 1986).

Evidence for migration takes a variety of forms. Birds may be observed migrating. For example, swallows, pipits and finches migrate during the day. Many species are strictly nocturnal migrants and are invisible though they may be audible.

Methodical observations at specific sites may reveal changes in abundance of partial migrants where only some individuals leave. Other species show peaks in spring and autumn consisting of birds on passage. Migration has also been observed by watching silhouettes of birds crossing the full-moon, or following shadows on radar screens or, in the case of a few individuals, by tracking birds with radio transmitters.

Although an experienced eye can discriminate between a migrating bird and one that happens to be flying over, one cannot judge the destination of migrants from one locality. Tying up peaks in different geographical areas may reveal breeding and wintering grounds and migration routes. However, the most satisfactory information comes from the recapture of marked birds. Extensive bird-ringing has been carried out in Europe for four decades and many migrant species have been recaptured in their wintering grounds. Some species travel no further than the Mediterranean, whereas others cross the Sahara Desert into central or even southern Africa. Southern United States and Central America are the principal destinations of migrants in North America. Southeast Asia, including Indonesia, and Australia are wintering grounds for many birds that breed in northern Asia (McClure 1974).

Why and How Do Birds Migrate?

Typically birds leave areas where food is declining or likely to become scarce. In the case of birds leaving polar or cool temperate areas in autumn the reasons for their departure are obvious. Not only will food be scarce for most birds but their energy requirements in winter will be high and the period of daylight in which to find food is short. Even though there are risks involved in migration those that leave have a greater chance of breeding in the following year. A few species do not migrate even from regions close to the Arctic Circle. They usually switch to foods such as seeds and berries, and some may set down their own stores. In the coldest areas the residents roost in sheltered places, like holes in trees, often communally.

If there are good reasons for leaving cold regions in autumn, why do these species leave the warmth and abundance of Central America, Africa and tropical Asia for the often unpredictable spring further north? There are probably two main reasons. First, temperate areas experience a flush of food, particularly insects, in spring and early summer, on which migrants can rear more young than in the less seasonal tropics. Second, individuals and species are packed tightly together in tropical areas, there is more intense competition for food and nest sites; in addition nest predators, particularly reptiles, are more common.

How birds migrate is a physiological problem outside the scope of this book. Typically they use changes in daylength as a cue for timing migration. The sun, stars and magnetism are all used to navigate, although the relative importance of these and how birds actually use them is not completely understood.

Bird Movements in Australia

Birds in the Australian region display the whole range of patterns of movements from total sedentariness (e.g., Marchant 1982) to intercontinental migration and extreme nomadism (Table 42 and Fullagar *et al.* 1988). The RAOU Atlas scheme has provided much

Table 42. Different categories of bird movements, with Australian examples.

Type of Movement	Species
Local Wandering ($<$ 1 km)	Fairy-wrens
Altitudinal Migration	Olive Whistler
Partial Migration	Silvereye
Intracontinental Migration	Yellow-faced Honeyeater
Intercontinental Migration	White-throated Needletail
Local Nomadism (1–100 km)	Lorikeets
Irruption outside normal range	White-browed Woodswallow
Continental Nomadism	Grey Teal

information on seasonal presences and absences, though all the data are yet to be fully analysed (Blakers *et al.* 1984).

Many non-migratory species gather into single-species or mixed flocks during the non-breeding season (see Chapter 4). These may wander outside the breeding territory in the case of Fairy-wrens and choughs. The co-operative thornbills occupy a clan territory which may include four or five breeding territories (see Figure 29). Brown Thornbills (*Acanthiza pusilla*) and Scarlet Robins (*Petroica multicolor*) tend to join flocks only as they pass through their breeding territories. Most of the honeyeaters are quite mobile outside the breeding season. Some of these movements involve local concentrations around flowering trees and shrubs. Others involve nomadism or regular migration and will be discussed later.

Altitudinal migration

Australia has few mountains and generally has mild winters, so consequently there are probably few altitudinal migrants (but see Tidemann *et al.* 1988). Olive Whistlers (*Pachycephala olivacea*) nest up to the tree-line in the Snowy Mountains and move down some 600–1,000 m to below the snow-line in winter (White 1986). This involves a migration of only a few kilometres. The populations of different valleys remain separate from each other. Flame Robins (*Petroica phoenicea*) also nest at high altitude. They migrate downhill and mostly westwards to the inland plains in autumn. Those in northern Victoria show some fidelity to their wintering sites (Fordyce 1978). Pied Currawongs (*Strepera graculina*) (Plate 41) also move out of higher forests in winter, tending to congregate in towns (Readshaw 1968).

A more unusual altitudinal migrant is the Eastern Spinebill (*Acanthorhynchus tenuirostris*) in northern New South Wales. It is abundant in upland heath in winter where banksia flowers well, but is scarce in other seasons, presumably moving to lower ground (McFarland and Ford 1987).

The Double-banded Plover (*Charadrius bicinctus*) and Wrybill (*Anarhynchus frontalis*) nest by upland streams in New Zealand and move down to the coast in late summer. About half of the population of the Double-banded Plover then migrates to winter in southeastern Australia (Lane 1987).

Intracontinental migration

Numerous Australian landbirds display regular and predictable movements. However, patterns of movements are often complex and in no case have they been fully explained (Fullagar *et al.* 1988). In addition, some populations of a species may migrate great distances whereas others may be sedentary. The Silvereye (*Zosterops lateralis*) (Plate 42) is one of the best studied examples of such partial migrants (Lane 1972; Rooke 1984). There are resident breeding populations in Sydney. These are augmented by other birds in late summer. In May large numbers of dark-flanked Tasmanian Silvereyes arrive and stay until September. This population is common in winter at least as far north as Armidale and Grafton.

Yellow-faced (*Lichenostomus chrysops*) (Plate 43) and White-naped Honeyeaters (*Melithreptus lunatus*) are perhaps the most conspicuous migrants in eastern Australia, migrating northwards in April and south again in August and September. They fly over in loose flocks just above the tree-tops, peaking in mid to late morning. Both species are less common in Victoria in winter (Paton 1980), yet increase in South Australia and northern New South Wales and southern Queensland (Hindwood 1956; Liddy 1966; Purchase 1985). Even so, some individuals could well be sedentary (Bell and Ford 1987). Scarlet Honeyeaters (*Myzomela sanguinolenta*) and Striated and Spotted Pardalotes (*Pardalotus striatus* and *P. punctatus*) are also regular migrants (V. Fazio, pers. comm.), though they are to some extent unpredictable in their movements. Their local abundance is probably related to the flowering of mistletoes and eucalypts and to the abundance of sap-sucking insects like scale and lerps.

Many of the flycatchers migrate along the east coast and ranges and Satin Flycatchers (*Myiagra cyanoleuca*) may cross both Bass and Torres Straits, though there is no evidence that individuals travel all the way from Tasmania to New Guinea. Other species that winter in northern Queensland and in New Guinea include White-throated Gerygones (*Gerygone olivacea*) (Plate 44), Rainbow Bee-eaters (*Merops ornatus*), Dollarbirds (*Eurystomus orientalis*), Sacred Kingfishers (*Halcyon sancta*) (Plate 45), Spangled Drongos (*Dicrurus hottentottus*) and Black-faced Monarchs (*Monarcha melanopsis*) (Blakers *et al.* 1984). Rainbow Bee-eaters often leave New Guinea in early spring against the southeast trade wind and may suffer mortality on the islands of the Torres Strait (Garnett 1985).

The Long-tailed Cuckoo (*Eudynamis taitensis*) and Shining Bronze-Cuckoo (*Chrysococyx lucidus*) leave New Zealand in autumn for the islands to the north, though the latter probably reaches eastern Australia in moderate numbers.

Intercontinental Migration

The only landbirds that migrate in large numbers between Asia and Australia are the Fork-tailed Swift (*Apus pacificus*) and White-throated Needletail (*Hirundapus caudacutus*). Both nest in eastern Asia including Japan and are in Australia from late October to March. New Guinea receives a few more regular migrants, including Grey and Yellow Wagtails (*Motacilla cinerea* and *M. flava*). The latter species and the Barn Swallow (*Hirundo rustica*) are regular though scarce visitors to northern Australia.

The paucity of such migrants is a reflection of the millions of years of isolation of Australia from Asia. Several native Australian birds that are presumably recent arrivals either belong to or are closely related to migratory species in Asia. These include the Bee-eater and Dollarbird, Richard's Pipit (*Anthus novaeseelandiae*), the Cisticolas, and the Bushlark (*Mirafra javanica*). The Pied Chat (*Saxicola caprata*) is apparently a migrant, which has very recently colonized New Guinea (Bell and Swainson 1985).

Two groups of birds display extensive migrations into and out of the Australian region from other continents. These are the seabirds and the waders.

Seabirds

Contrary to the usual trend of increased diversity of species towards the equator, the seabirds are most diverse in temperate and sub-polar regions. A sizeable proportion of the world's seabirds nest south of Australia and are regular or erratic visitors in the non-breeding season. Many albatrosses, including the great Wandering Albatross (*Diomedea exulans*), which nests in the Crozet and Kerguelen Islands between Africa and Antarctica and other sub-Antarctic islands, fly around the world in the westerly wind belt (Weimerskirch *et al*. 1985). Birds banded elsewhere have been frequently recovered off Australia, especially off the New South Wales coast and Australian banded birds have been recovered in breeding colonies. Giant Petrels (*Macronectes*), which nest on the Antarctic mainland as well as sub-Antarctic islands also occur off Australia and New Zealand. Many other Antarctic seabirds have been banded without showing very many recoveries yet, so there is still a great deal to be learned about their movements (Weimerskirch *et al*. 1985).

Australian breeding seabirds also display extensive migrations and the best known of these is the Short-tailed Shearwater or Mutton-bird (*Puffinus tenuirostris*). This bird breeds in millions in the Bass Strait (see Chapter 8), and departs in May at first eastwards towards New Zealand (Figure 36). It then flies due north to be off Japan in June. By July large numbers are south and west of Alaska and in August they are streaming back across the equator to return to their breeding colonies by late September. Other shearwaters show similar movements but do not travel as far. The Wedge-tailed Shearwater (*P. pacificus*) (Plate 46), for instance, spends the winter off the Philippines.

The terns are also great travellers though most species that breed in Australia tend to disperse widely rather than migrate.

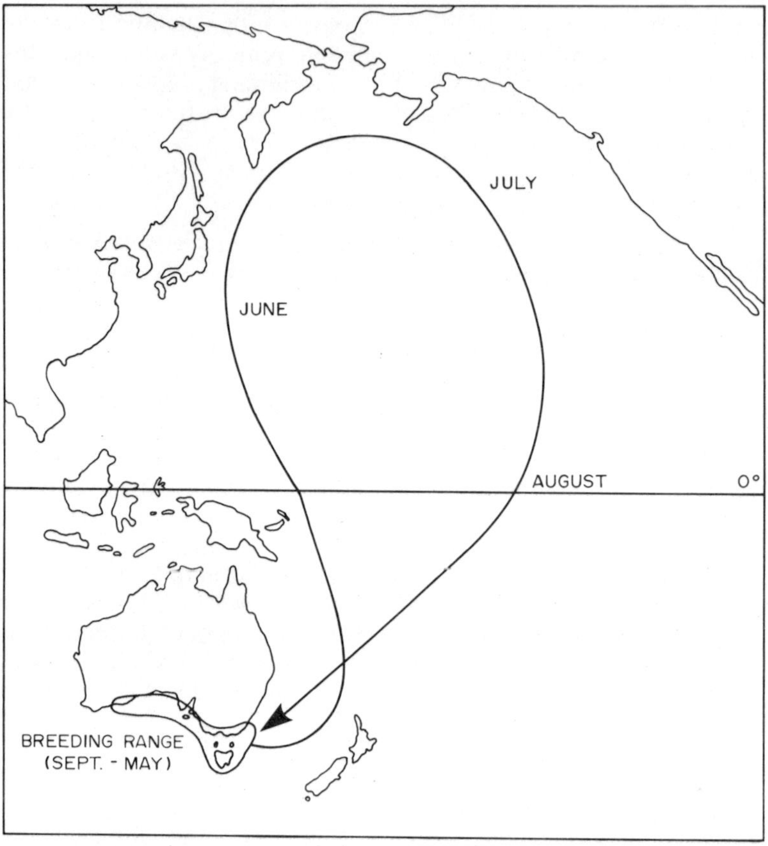

Fig. 36. Migration route of the Short-tailed Shearwater (*Puffinus tenuirostris*) (modified from Rowley 1974). Note that birds move in a broad front rather than along a narrow route.

The Common Tern (*Sterna hirundo*) breeds in temperate Europe and Asia and reaches Australia in reasonable numbers in the southern summer. The Arctic Tern breeds further north and it too, spends the non-breeding season even farther south. Arctic Terns banded in Scandinavia and the USSR have been recovered in New South Wales and Western Australia. The Arctic Skua or Jaeger (*Stercorarius parasiticus*) is another bird that breeds in the far north, often preying on lemmings. It pursues Crested Terns (*Sterna bergii*) off the east coast of Australia from November to March.

Waders

The last decade has seen a great increase in knowledge about waders or shorebirds, especially the non-breeding migrants (Lane 1987). About two-thirds of the regularly occurring species in Australia are visitors from the Palaearctic or Nearctic. The more common species are listed in Table 43. Waders occur in large numbers in a few sites in southern Australia, especially near Adelaide and Melbourne (Figure 37). It has been discovered only recently that vast numbers also occur in northern Australia, in particular at Eighty Mile Beach, Roebuck Bay and in the Gulf of Carpentaria (Plate 27).

Table 43. The more abundant migrant waders that visit Australia (from Lane 1987).

Species	Breeding area	Main Australian "winter" area
Grey Plover *Pluvialis squatarola*	High arctic, Asia and N. America	Mostly western and southern coasts
Large Sand Plover *Charadrius leschenaultii*	Gobi Desert, Mongolia, China	Northwestern and northern Australia
Eastern Curlew *Numenius madagascariensis*	Eastern Siberia, Mongolia	Eastern Australia
Little Curlew *N. minutus*	Siberia	Grassland in northern Australia
Black-tailed Godwit *Limosa limosa*	Temperate Europe and Asia	Northern Territory and Gulf of Carpentaria
Bar-tailed Godwit *L. lapponica*	Arctic Siberia and Alaska	Northwestern and eastern coasts (also New Zealand)
Great Knot *Calidris tenuirostris*	Northeastern Siberia	Northern Australia
Red Knot *C. canutus*	High arctic, Siberia and North America	Most coasts
Sharp-tailed Sandpiper *C. acuminata*	Northeastern Siberia	Northwest, southeast and inland.
Red-necked Stint *C. ruficollis*	Northern Siberia and Alaska	Northwest and southeast
Curlew Sandpiper *C. ferruginea*	Northern Siberia	Northwest and southeast

The regular migrants range from the Red-necked Stint (*Calidris ruficollis*) at 30 grams to the Eastern Curlew (*Numenius madagascariensis*) weighing nearly a kilogram. The breeding

Fig. 37. Major sites for migratory waders in Australia (from Lane 1987).

grounds of a selection of species that winter in Australia are shown in Figure 38. Most of the migrants arrive in northern Australia in late August and September. Many, in particular Large Sand Plovers (*Charadrius leschenaultii*), and Great Knots (*Calidris tenuirostris*), remain there during the non-breeding season. Others, such as Red-necked Stints and Curlew Sandpipers (*Calidris ferruginea*), continue southwards. Those in the southeast start arriving in October and build up to a peak around New Year. Most depart in March from the south and in April from the north. They will reach their breeding grounds in late May or early June and some start to return as early as July. In many species first-year birds remain in Australia through the winter. Bar-tailed Godwits (*Limosa lapponica*) and Red Knots (*Calidris canutus*) move on to New Zealand in large numbers, as do a few of the other species (Sagar 1986).

Although most waders occur on estuaries or mudflats a few prefer other habitats. Little Curlews (*N. minutus*) occupy grasslands across northern Australia, moving inland when the wet season arrives. Latham's Snipe (*Gallinago hardwickii*), the entire population of which migrates from Japan to Australia, prefer freshwater swamps, mainly in the southeast. Small numbers of the typically coastal species also occur regularly on inland waters, in particular Sharp-tailed Sandpipers (*Calidris acuminata*). In very wet years the large inland salt lakes, such as Lake Eyre, hold vast numbers of migratory waders.

A variety of methods has been used to unravel the mysteries of the destination and timing of migration in waders. Counting at important sites has been valuable. Banding has often produced spectacular results, especially now that the Republic of China has opened up communication with the outside world (Weishu and Purchase 1987). In addition, dyeing of waders has allowed identification of individuals, without the necessity to recapture birds, and coloured leg flags are now being used. Captured birds can also provide valuable clues. Different breeding populations may differ in beak or wing-length. Examination of the wings and tail for moult can help to separate young from adults and may indicate whether migration is imminent. Also waders, like other migrants, put on extensive deposits of fat before departing. Finally, radar can be used to determine the direction and speed of departing flocks.

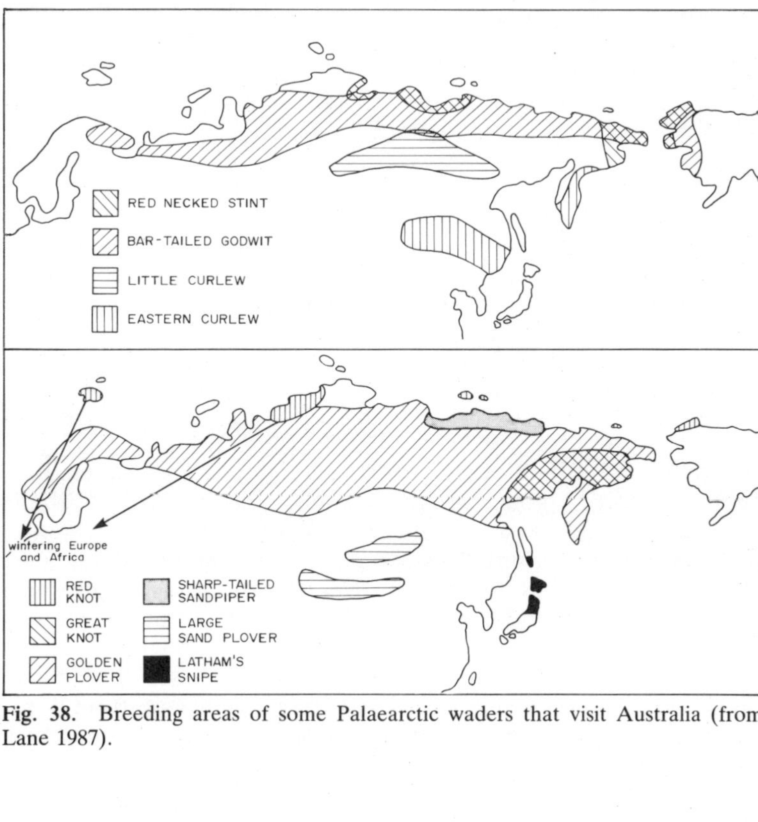

Fig. 38. Breeding areas of some Palaearctic waders that visit Australia (from Lane 1987).

The destinations of waders leaving from northwestern Australia have been worked out by a combination of methods (Lane and Jessop 1985). First, birds were captured and weighed on Eighty-Mile Beach near Broome. Second, flocks departing in the evening were noted and then followed on radar at Broome airport. This allowed direction and speed to be calculated. This information was then related to knowledge of where each species occurred as a passage migrant in Southeast Asia. Birds left in a northwest to NNW direction at about 60–90 km per hour, often with a tailwind. The tiny Stints carry enough fat to cover about 2,400 km at this speed, which would bring them easily to the south coast of Borneo. The slightly larger Curlew and Terek Sandpipers (*Tringa terek*) could reach as far as western Malaysia and Thailand, whereas Turnstones (*Arenaria interpres*) and Tattlers (*Tringa brevipes*) could fly over 4,000 km to the southern coast of China. As premigratory fattening is common in waders it seems that they may typically complete their migration in a series of four or five long flights. Quite likely many waders cross from southeastern Australia to the north coast non-stop. On the southern migration Curlew Sandpipers on the Malay Peninsula accumulate enough fat to reach the north coast of Australia. Red Knots migrate from South Africa to Siberia in three or four hops, travelling possibly as far as 6,900 km in one flight (Dick *et al.* 1987).

Waders are indeed remarkable intercontinental migrants and future work on their movements will require extensive international co-operation, which is developing. There is now extensive collaboration between observers in Australia and Southeast Asia, and reciprocal migratory bird legislation has been developed or is being developed between Australia and several Asian countries.

Nomadism

Whereas true migration is regular and predictable, nomadism implies unpredictable, even random, movements. Perhaps migrants respond to cues such as daylength or internal circannual rhythms, whereas nomads respond to local weather and food supply. Keast (1961) has argued that nomadism is an important feature of Australian birds, especially those that breed in the dry centre of the country. However, as mentioned in the introduction, one should make a distinction between species that show fairly regular seasonal movements and may irrupt elsewhere, from those whose movements are highly unpredictable. Obviously migration and nomadism are at the ends of a continuum and many species fall between these extremes.

The Budgerigar (*Melopsittacus undulatus*) is one of the most abundant birds of inland Australia, in some years occurring in vast flocks. In other years it may be scarce in or absent from large areas of southern Australia and has consequently been regarded as a nomad. Wyndham (1982) however, showed that there was a regular north–south pattern to the Budgerigar's movements. It occurred in the south mostly in spring and summer but retreated to the north in autumn and winter. It does change greatly in abundance from year to year and in the exact sites where it settles. The Black Honeyeater (*Certhionyx niger*) is another species often classed as a nomad. Again, there is a regular north–south movement, though the species is very scarce in many years (Ford 1978). White-browed Woodswallows (*Artamus superciliosus*) (Plate 47) are typically irruptive, often breeding in large numbers outside their normal range (Recher and Shultz 1983). Their appearances there though are almost invariably in late spring and they usually depart before the end of summer. Crimson Chats (*Ephthianura tricolor*) also irrupt from their inland range to the south and east in spring and summer and perhaps to the southwest in winter and spring (Keast 1958b; Blakers *et al.* 1984). They appear to move less regularly and more erratically than the previous species, so are close to being true nomads.

Emus (*Dromaius novaehollandiae*) are great wanderers and have now been well studied in southwestern Australia (Davies 1984). They tend to move south and towards the coast in autumn and winter and northwards and inland in spring and summer. This overall regularity can be upset though when the weather is atypical. Summer rain tends to be more frequent in the north and winter rain in the south. In one year though rain fell in the south in February and Emus moved south and subsequently north again two months later. Emus can be regarded either as rather regular migrants or as nomads that respond to conditions locally or a few hundred kilometres distant. Davies thought that birds may move towards cyclonic banks of cloud or possibly thunder and lightning, which can be detected over a hundred kilometres away.

Just as in northern temperate regions, Australian raptors that specialize on rodents are nomadic. The most common species are Black-shouldered Kites (*Elanus notatus*), Kestrels (*Falco cenchroides*) and Brown Falcons (*F. berigora*), though the less specialized Black (*Milvus migrans*) and Whistling Kites (*Haliastur sphenurus*) may also fluctuate greatly in numbers. Letter-winged Kites (*Elanus scriptus*) provide an extreme case, building up in the centre when the Plague Rat (*Rattus villosissimus*) is numerous, then dispersing widely when the rat outbreak finally crashes (Jackson 1919; Hollands 1979).

One of the most nomadic species is the Flock Bronzewing (*Phaps histrionica*), which is perhaps the Australian equivalent to Pallas's Sand-grouse. The core of its population occurs in the grassy plains of the Barkly Tableland in Northern Territory (Frith 1982). From this region it irrupts throughout most of inland northern Australia. Infrequently irruptions may spread farther south. Gould encountered large numbers on the northwestern slopes of New South Wales, where it is virtually never seen today. Whereas it could have been resident there it seems more likely that these birds resulted from a periodic invasion. Stubble Quail (*Coturnix pectoralis*) are also nomadic in southeastern Australia as banded birds have been found to disperse in all directions (Frith and Waterman 1977).

Many landbirds of inland Australia show regular to irregular movements in response to seasonal or year-to-year changes in conditions, but it is the waterbirds that are most markedly nomadic. Frith (1962b) showed that Grey Teal (*Anas gibberifrons*) (Plate 48) dispersed in all directions as inland swamps dried out. The Black-tailed Native-hen (*Gallinula ventralis*) occasionally irrupts into the Adelaide area in vast numbers (Matheson 1978). A notable invasion occurred in 1975 after two years of good rain in northeastern South Australia. After heavy rain the inland rivers and salt lakes fill with water, and pelicans, cormorants, ducks, terns, gulls and waders arrive, sometimes in vast numbers. Lake Eyre was teeming with waterbirds, including 95,000 Red-necked Avocets (*Recurvirostra novaehollandiae*) in 1984. Most of these species breed also at a wide range of more coastal sites in drier years. An exception is the Banded Stilt (*Cladorhynchus leucocephalus*). This species breeds in vast numbers on inland salt lakes in South Australia and Western Australia (McGilp and Morgan 1931; Burbidge and Fuller 1982), but at infrequent intervals. Repeated breeding attempts are made until the water level declines when the colonies are deserted, leaving thousands of eggs and young birds to die. The Stilts then retreat to coastal sites, especially saltfields and other areas of high salinity until the next heavy fall of rain over the inland, which may be several years' time. Probably only a few of the breeding sites are known and these may all be used irregularly.

Summary

In conclusion, Australasia is the destination of numerous species of those cosmopolitan travellers, the waders and seabirds. Very few landbirds cross the continental boundaries, however. Two swifts from Asia reach Australia in large numbers and a handful of other species spend their non-breeding season in New Guinea and trickle into Australia. These do not compare with the massive invasions of Africa and Central America by birds from the north. Although many Australian birds winter in New Guinea only a few go as far as Indonesia, virtually none reaches the Asian mainland. A few New Zealand birds migrate to Australia or the islands of the western Pacific.

Within Australia there are extensive movements of birds, usually over relatively short distances and often following complex patterns. Few species are total, regular migrants except in the southeast. Many are partial migrants or somewhat unpredictable in their movements. Probably the incidence of random nomadic movements has been exaggerated though. At the moment little is known about the details of the migration of any of these species. Nix (1976), by using his model of seasonal changes in plant growth, predicted where migratory birds should go at different times of the year. Presences and absences at specific sites give his predictions some support, but there is a need for national monitoring of seasonal changes in abundance of birds. Bird-banding has so far provided only sketchy information on the travels of individual birds. Where populations differ in plumage or measurements, as in the Silvereye, detailed examination of captured birds may reveal something of their origins or destinations (for instance Rainbow Bee-eaters captured in the Torres Strait have measurements similar to those breeding in southeastern Australia). Perhaps in such a vast and sparsely populated country even a great increase in bird-banding may not increase knowledge much. Better co-ordination of banding effort as is now taking place in the Australian Bird and Bat-Banding Scheme and the establishment of Bird Observatories by the RAOU will certainly help. More bird-banders are desperately needed. Very few people have actually recorded migration taking place, though this is not hard for diurnal migrant honeyeaters, pardalotes and swallows. Perhaps other techniques should be developed. Dyeing has already been used in waders and some landbirds (e.g., Yellow-faced Honeyeater — K. Lowe, pers. comm.); it could be used in other waterbirds or landbirds. Where species are highly concentrated they could be bombed with fluorescent dyes, a method that was successful with Queleas (Jaeger *et al.* 1986). Is there any scope for measuring rare elements in feathers to identify sites where moulting may have taken place? Could food be loaded with radio-active isotopes? Maybe a Geiger Counter could become a part of any bird-bander's equipment. Song Thrushes (*Turdus philomelos*) wintering in Spain contained radionuclides in 1986 which they probably picked up in Fenno-scandia, following the Chernobyl accident (Ruiz *et al.* 1988). Could birds feeding near radioactive deposits and mines in northern Australia contain unique frequencies of radionuclides, which could be detected if they were captured elsewhere?

There is no doubt that progress will be made using traditional methods especially if more bird-watchers are drawn in to help and all is coordinated well. Rapid advances will follow the use of original methods though. There is some urgency in finding out the main migratory routes of landbirds as so much habitat has been or is being lost or degraded. For instance, it is not known what effect the extensive clearing of brigalow in central Queensland might have had or will have on birds nesting in southeastern forests and woodlands. Australian landbirds appear to migrate in short hops rather than in long hauls after putting on fat. The conservation of populations therefore requires a chain of reserves along migratory routes. The next decade should see exciting developments in the knowledge of the migration of birds within Australia.

10. The Ecology of Rare, Endangered and Extinct Birds

Birds and people have influenced each other, both positively and negatively, for many thousands of years, but the effect of man on birds has expanded and accelerated in recent centuries. People have enjoyed the brilliant colours and attractive songs of birds and they have envied the freedom that flight has allowed them. Human cultures have deified birds. Poetry and literature have frequently included birds, often in an idealistic, romantic or allegorical way. Birds have appeared on coats-of-arms and flags. Fledgling conservation bodies on both sides of the Atlantic have usually focussed on birds. The most obvious negative effect of people has been to reduce the population sizes of birds, sometimes to extinction. This can be due to straightforward biological processes like predation and competition, but also recently from technological changes that have led to habitat destruction and pollution. Some species of birds have benefited from human activity, notably those that have been able to adapt to the simple habitats that have usually been produced, or to new environments that have been created. A few birds have become major competitors for food, especially grain. Very recently some of the species that were most adversely affected by humans have been protected and actively conserved in modified habitats, even in captivity.

In this chapter I shall first discuss the types of birds worldwide that have become endangered or extinct in recent centuries. Then I shall look in detail at the very different situations in Australia, New Zealand and other islands in the Pacific.

Overall I believe that there is no reason to be pessimistic about the future of birds in the region, providing that there is a response to the desperate need for financial support and dedicated scientists and managers.

Extinction

Many large mammals on the world's continents became extinct

during the Pleistocene, coincident with the arrival of man. Only Africa has retained a reasonably intact megafauna probably because it coevolved with man's increasing efficiency and mastery over the environment. Several large flightless birds also became extinct in prehistoric times and one can only presume that smaller species, which left fewer fossils, were also lost.

Since 1600 about 164 species or subspecies of birds have become extinct, though a few may yet be re-discovered. The Red Data Book lists 318 taxa of birds as currently under threat of extinction, some of which are possibly extinct (King 1981). Amongst these there is a great predominance of species inhabiting forests (Table 44) and over half of all endangered species are found on islands. The major cause of endangerment is habitat

Table 44. Percentage of endangered species of birds in the world that occur in different habitats (King 1977).

Forest	64%
Wetland	12.7%
Grassland	7%
Scrub	9.8%
Others	6.5%

destruction, typically the clearing of forest, though next in importance is the draining or pollution of wetlands (King 1977). Direct killing by people or introduced predators is also important.

Endangered birds are a heterogeneous collection with respect to morphology and ecology, but many of the species are rather large (Halliday 1980b). Such birds tend to occur at low density, to require large territories and have low reproductive rates ("K"-selected – see Chapter 5). Raptors are typical examples, as they are at the top of the food chain and are naturally scarce. They are vulnerable to the concentration of pesticides like the chlorinated hydrocarbons. Large birds also provide attractive targets for shooters, and even the Californian Condor (*Gymnogyps californicus*), one of the rarest species of birds in the world, was occasionally shot during its last decade of freedom.

As mentioned above, forest and island birds are two groups that appear frequently on the list of threatened species. Both live in what used to be relatively stable environments, and accordingly tend to have high adult survival rates but low reproductive rates. Forest birds are often rather specialized and so are less able to cope with disturbance of their habitat. Disturbance often takes the form of total clearance, with which the forest specialists cannot cope. Birds of drier woodlands, which experience a rather unpredictable environment, with drought and fire, seem better able to cope with extensive habitat disturbance. By contrast, island birds are often generalists but they are adapted to rather atypical environments with unique plants and an absence of predatory mammals. Replacement of native habitats with monocultures of continental origin and the introduction of rodents and carnivorous mammals have all too often been too much for island endemics throughout the world.

Another distinct group that makes up a significant proportion of endangered or extinct birds is the colonial seabirds. For a start most live on islands, but they are particularly vulnerable because they concentrate into a few sites where they occur at high density. Again they have low reproductive rates and very high rates of adult survival (see Chapter 8). Any factor that consistently reduces their reproductive rate or increases adult mortality will tend to lead to a steady decline. It may take many years for population changes to be recognized, because there is often a non-breeding component to healthy populations and because seabirds often do not breed until several years old. Losses to a colony can be replaced by the previously non-breeding birds and by birds starting to breed at an earlier age. Direct exploitation by man has led to the loss of some species, notably the Great Auk (*Alca impennis*) of the North Atlantic. Cats and rats are more often the exterminating agent. Many populations of seabirds are being or have been exploited by man without disastrous effects, for instance the Short-tailed Shearwater (*Puffinus tenuirostris*) in Bass Strait. Others have been reduced but with protection have recovered, as in the case of many seabirds around the North Sea.

Extinct and Endangered Birds in Australia

Aboriginal man arrived in Australia between 100,000 and 40,000 years ago. Many large mammals and some species of large flightless birds have disappeared since that time. Whereas it is likely that hunting and habitat alteration, especially fires, may have caused the decline of these species, climatic changes, in particular arid periods, could also have contributed.

The 200 years of European settlement in the Australian region have seen further and often dramatic changes in the avifauna. Generally, the Australian mainland has fared much better than the surrounding islands. Probably only a single mainland species, the Paradise Parrot, has gone extinct. Paradise Parrots bred in termite mounds in eucalypt woodland, mostly to the west of the Great Dividing Range from northern New South Wales to northern Queensland (Blakers *et al*. 1984). This area has suffered extensive clearing and heavy grazing and also has large numbers of feral animals like pigs and foxes. However, many other parrots in this region have fared well since the arrival of Europeans and it is unclear why Paradise Parrots suffered so badly.

Numerous species have declined since European settlement, and several of these may be endangered or at least vulnerable. Our knowledge of many rare species is insufficient to judge how endangered they are but a list of threatened species is given in Table 45. The worldwide tendency for endangered birds to be

Table 45. Extinct, threatened and rare species of birds found in Australia and its nearby territories. (Habitat: RF – rainforest, F – forest, W – woodland, S – semi-arid scrub, H – heath, G – grassland, D – desert, FW – freshwater, M – marine). Based on *List of rare, endangered and extinct bird species of Australia and its island territories*. RAOU Technical Report, *in prep*. J. Brouwer and S. Bennett (eds.).

Species	Habitat
Extinct	
King Island Emu *Dromaius ater*	W
Kangaroo Island Emu *D. baudinianus*	W
Possibly Extinct	
Paradise Parrot *Psephotus pulcherrimus*	W
Endangered	
Abbott's Booby *Sula abbotti*	M
Christmas Island Frigatebird *Fregata andrewsi*	M
Black-bellied Button-quail *Turnix melanogaster*	RF
Orange-bellied Parrot *Neophema chrysogaster*	H
Noisy Scrub-bird *Atrichornis clamosus*	F
Black-eared Miner *Manorina melanotis*	S
Norfolk Island Silvereye *Zosterops albogularis*	F
Goudian Finch *Erythura gouldiae*	W
Vulnerable	
Malleefowl *Leipoa ocellata*	S
Lord Howe Island Woodhen *Tricholimnas sylvestris*	F
Plains-wanderer *Pedionomus torquatus*	G
Hooded Plover *Charadrius rubicollis*	M
Golden-shouldered Parrot *Psephotus chrysopterygius*	W
Hooded Parrot *Ps. dissimilis*	W
Carpentarian Grasswren *Amytornis dorothiae*	G
Western Bristlebird *Dasyornis longirostris*	H
Forty-spotted Paradalote *Pardalotus quadragintus*	F

Table 45 — continued

Rare (including those whose status is poorly known)	
Freckled Duck *Stictonetta naevosa*	FW
Red Goshawk *Erythrotriorchus radiatus*	F
Grey Falcon *Falco hypoleucos*	S
Buff-breasted Button-quail *Turnix olivei*	F
Cox's Sandpiper *Calidris paramelanotos*	M
Palm Cockatoo *Probosciger aterrimus*	RF-W
Glossy Black-Cockatoo *Calyptorhynchus lathami*	W
Baudin's Black-Cockatoo *C. baudinii*	F
Eclectus Parrot *Eclectus roratus*	RF
Alexandra's Parrot *Polytelis alexandrae*	D
Ground Parrot *Pezoporus wallicus*	H
Night Parrot *Geopsittacus occidentalis*	D
Scarlet-chested Parrot *Neophema splendida*	S
Rufous Owl *Ninox rufa*	RF
Sooty Owl *Tyto tenebricosa*	RF
Lesser Sooty Owl *T. multipunctata*	RF
Albert's Lyrebird *Menura alberti*	F
Rufous Scrub-bird *Atrichornis rufescens*	F
Red-lored Whistler *Pachycephala rufogularis*	S
Western Whipbird *Psophodes nigrogularis*	H
Purple-crowned Fairy-Wren *Malurus coronatus*	W
Mallee Emu-wren *Stipiturus mallee*	S
Black Grasswren *Amytornis housei*	G
White-throated Grasswren *A. woodwardii*	G
Eyrean Grasswren *A. goyderi*	G
Grey Grasswren *A. barbatus*	G
Thick-billed Grasswren *A. textilis*	D
Eastern Bristlebird *Dasyornis brachypterus*	S
Chestnut-breasted Whiteface *Aphelocephala pectoralis*	D
Eungella Honeyeater *Meliphaga hindwoodi*	RF
Regent Honeyeater *Xanthomyza phrygia*	F-W
Yellow Chat *Ephthianura crocea*	G
Yellow-rumped Mannikin *Lonchura gouldiae*	G

inhabitants of forests (Halliday 1980b) is not true of Australia where forest communities have survived rather better than those of semi-arid woodlands and scrubs. Only a few forest birds can be considered endangered. The Noisy and Rufous Scrub-birds are relict species, inhabiting the understorey of moist forests (G. T. Smith 1985; Ferrier 1985). The ranges of both have contracted over the last century. The Noisy Scrub-bird occurred for a long period on only one coastal patch of scrubby forest in southwestern Australia. It perhaps survived there because the area was protected from the frequent and extensive fires lit by Europeans (Smith 1979a, 1985). The scrub-birds provide a curious paradox. They have very poor powers of dispersal and low reproductive rates. Yet they appear to have occurred as small isolated populations. Rufous Scrub-birds frequent the edges of temperate rainforests or occupy patches within subtropical rainforests where there have been disturbances. Noisy Scrub-birds occur in swamp-forest *ecotones*. Such patches are usually small, rather unstable and may be ephemeral. The scrub-birds' life-style therefore does not seem to adapt them very well to their favoured habitats. Although the range of the Rufous Scrub-bird has retreated, especially at lower altitude, several healthy populations still exist. Noisy Scrub-birds have been studied for 20 years, during which their population has increased. They have recently been introduced to a second site. Forest-dwelling birds like black-cockatoos and owls, which require hollow trees for nesting, may be threatened by logging and clearing of forests.

Many threatened birds in Australia inhabit low scrub of one type or another. The Ground Parrot (*Pezoporus wallicus*) occurs in low heathlands and sedgelands from southeastern Queensland to Victoria, and also in southwestern Western Australia and Tasmania (Meredith 1984; Meredith *et al.* 1984). Its range has become reduced and fragmented since European settlement. The principal reasons are clearance of habitat for grazing, an altered frequency of fires, and predation by introduced mammals. Heathlands are unsuitable for Ground Parrots for a few years after a major fire, then become more suitable until 20–25 years after the fire. Ground Parrots are unable to survive in regions where fires are annual, occur every few years or have not occurred for 20 years. Generally, Ground Parrots are still widespread and relatively common in parts of Tasmania and eastern Australia, though the Western Australian subspecies is scarce. Many populations are in reserved areas, and management of these to favour Ground Parrots is possible.

The same optimistic state of affairs cannot be claimed for the Orange-bellied Parrot. This attractive species is a migrant (see Chapter 9) meaning that two areas and several habitats have to be protected and managed, in three states. There are fewer than 200 birds left, possibly not many more than 100 (Brown *et al.* 1985). The breeding areas are rather extensive, and relatively undisturbed, though again the frequency of fire may be critical. Breeding success is reasonably high. The species appears to be limited more by its wintering habitat. This is coastal saltmarsh in Victoria and dunes in South Australia. The Victorian habitat has been greatly modified by industrial and other developments. That in South Australia experiences disturbance from off-road vehicles and modification by grazing. Protection of winter habitat, at present in many hands, is clearly important, as is the minimization of disturbance by people. The species' wintering habitat near Melbourne is particularly under threat.

Another parrot, the Night Parrot, remains an enigma. It inhabits the vast spinifex (*Triodia*) plains of the inland, which have been used little by Europeans. They have been grazed by cattle and sheep at very low density and perhaps more importantly by feral animals. Night Parrots were reported frequently in the 19th century but have only been seen occasionally over the last 50 years. It is unclear whether this is because they are very rare or simply very hard to observe. It is hard to see what could be done to help the species short of removing grazing by domestic stock, and the control of feral animals, a very difficult task.

Parrots generally, and also some species of finches, are illegally trapped and exported as pets. Trapping presents a major threat to the rarer species (e.g., Gouldian Finch) and a potential threat to others. The necessarily strict laws on export of birds need to be stringently enforced.

Several raptors are very rare in Australia, for instance the Red Goshawk and Grey Falcon. They were probably never common but have been recorded over a wide area of Australia. They do not appear to live in areas of heavy use of pesticides, so chlorinated hydrocarbons, which have devastated some raptor populations in Europe and North America, do not appear to be to blame (Newton 1979). Red Goshawks appear to require mature trees near water for breeding and this habitat has deteriorated in northern Australia, due to overgrazing and frequent burning (Debus and Czechura 1988). Also the Red Goshawk's main prey species, waterbirds, pigeons, parrots and large passerines, may have declined. It is important to find out more about the ecology of Australian raptors so that if populations decline this can be detected, the reasons for it discovered and the problem remedied.

Waterbirds have suffered the loss and disturbance of habitat in an already dry continent. Only one species gives cause for concern at present: the Freckled Duck (Martindale 1986). Freckled Ducks are anatomically rather unusual and possibly ancestral to other ducks (Frith 1967). They are widespread when not breeding, in southeastern and southwestern Australia. They appear to converge on the swamps of northwestern New South Wales and southwestern Queensland to breed, though little is known about this aspect of their biology. Most breeding takes place in spring, perhaps mostly in years of good inland rains. Clutches average seven eggs and range from five to fourteen. Freckled Ducks clearly have a high potential reproductive rate, but chance factors like the weather and also damage to swamps by grazing animals may disrupt or prevent breeding. In very dry years Freckled Ducks retreat to a few coastal wetlands. After the severe drought of the early 1980s about 8,000 birds were counted, though some, perhaps many, would have been missed. This reasonable number along with the high reproductive rate do not seem to make the Freckled Duck a particularly endangered species. However, despite protection, hundreds may be shot in duck-hunting seasons, when inland swamps dry out and ducks are concentrated near population centres (Martindale 1986). Freckled Ducks tend to concentrate in a few sites and are easily mistaken for more common unprotected species so most shooting is unintentional. The main problem with the Freckled Duck at present is lack of knowledge of its breeding areas and general biology. If the former is protected and hunting banned on wetlands in times when they hold large numbers of Freckled Ducks, then there is no reason why Freckled Ducks should become rarer. Worldwide, waterfowl have frequently been rescued from near extinction, mainly by captive breeding. This last resort should not be required for the Freckled Duck.

Overall the avifauna of the Australian continent appears to be in a healthy state. Many of the more obviously threatened species are being studied by State national parks and wildlife services, the CSIRO or RAOU, or in projects supported by the World Wildlife Fund, despite the fact that these organizations have to struggle for meagre funds. One should not become complacent however as species can slip away almost unnoticed. The Black-eared Miner, an inhabitant of dense mallee on the Victorian, South Australian and New South Wales borders is one such species. Clearing of the mallee has allowed the spread of the Yellow-throated Miner (*Manorina flavigula*), which can hybridize with the Black-eared Miner (Joseph 1986; Considine 1986). Only a handful of pure Black-eared Miners may remain. The Regent Honeyeater was common enough in the last century to be regarded as a pest in orchards. It is now rarely seen and is perhaps thinly spread over a wide area (Franklin *et al.* 1987). In contrast, the Helmeted Honeyeater (*Lichenostomus m. cassidix*), a well differentiated subspecies of the Yellow-tufted Honeyeater (*Lichenostomus melanops*), is restricted to a few valleys northeast of Melbourne (Wykes 1985). Habitat loss is the basic reason for its decline, but the few remaining colonies may be threatened by competition with Bell Miners (*Manorina melanophrys*). Interspecific interactions may be a factor in the decline of all three of these species. In these circumstances management is far harder to implement.

Tasmania has not so far suffered the extinctions that many other islands have, possibly because the climate has improved since it was isolated from the mainland, and it still has an unsaturated avifauna. Only one of its endemic species, the Forty-spotted Pardalote, is declining (Woinarski and Bulman 1985), though the very rare Orange-bellied Parrot breeds only in Tasmania.

Extinct and Endangered Birds in New Zealand

In contrast to the situation in Australia, birds in New Zealand have suffered severely from the effects of man (Mills and Williams 1979). Polynesians settled New Zealand about 1,000 years ago and had a devastating effect on the birds, as between 34 and 45 species became extinct. The most impressive of these were the moas, relatives of emus and ostriches, which included the tallest of birds. They were hunted by the Maoris, who also burned much of the original vegetation. The Kiore (*Rattus exulans*) and Polynesian dog (*Canis familiaris*) were also introduced and probably accounted for the loss of flightless rails and some other birds. Seven species of waterfowl disappeared and six rails, along with two hawks and two eagles.

Europeans have cleared large areas of forest, especially on the North Island and the eastern side of South Island and a further eight species and seven subspecies have become extinct. These included members of three endemic families – both populations of the Piopio or thrush (*Turnagra capensis* – Turnagridae), the Huia *(Heterolocha acutirostris)* and probably all populations of the Bush Wren (*Xenicus*). The origin of these families is certainly ancient and their relationships unclear though the last has its closest relatives in South America (Sibley *et al.* 1982). Birds on the offshore islands, including the distant Chatham Islands have suffered badly, losing many races and a few species. On the other hand small islands have also provided refuges for species beleaguered on the mainland. Two species that presumably colonized from the Northern Hemisphere are amongst the losses: the endemic Snipe (*Coenocorypha aucklandica*) and the Auckland Island Merganser (*Mergus australis*).

Clearance of habitat was undoubtedly a major contributor to extinction of many species but perhaps the main reason was deterioration of the remaining habitat by introduced herbivores and predation on eggs, nestlings and adults by introduced predatory mammals. Goats, deer and possums have damaged the forests and grasslands and introduced predators include rats, mustelids and cats. In addition to the species that were exterminated, a further six species are considered endangered and 11 more rare by Mills and Williams (1979). Some of these are remnants of species whose mainland populations have become extinct.

There are too many species to treat individually, but there are some patterns amongst species that are now most endangered and those whose populations are reasonably stable. Sadly, the very rare species include many of the more remarkable, even bizarre forms, which belong to the endemic families or genera. Such species require large areas of natural habitat.

The Kakapo (*Strigops habroptilus*) is a huge, flightless parrot which is nocturnal and in which the males boom from a depression in the ground to attract females (Merton *et al.* 1984). It was originally plentiful throughout New Zealand but has declined steadily until only tiny remnants of populations remain in Fiordland and Stewart Island in the extreme south. Its unusual social behaviour and other habits, and the fact that males fight vigorously, make it a difficult species to conserve in the wild or breed in captivity.

The Takahe or Notornis (*Notornis mantelli*) is a giant flightless rail that was considered extinct until rediscovered in 1948. It had a large pre-historic range but now only occurs in a small area of alpine grassland in Fiordland. It may have declined because its habitat, tussock grassland, was largely replaced by forest after the last ice age (Mills *et al.* 1988). Alternatively, it could have inhabited forest until the Maoris arrived, who with their hunting, and the dogs and rats they introduced, restricted it to the extreme southwest of South Island (Beauchamp and Worthy 1988). It has suffered recently from competition with introduced deer, though it had increased to 180 individuals by 1986, after the deer had been reduced (Mills *et al.* 1988).

The Kokako (*Callaeas cinerea*) and Saddleback (*Philesturnus carunculatus*) survive as small isolated populations, the latter on offshore islands. They are the only survivors of the wattlebird family (Callaeidae). The Stitchbird (*Notiomystis cincta*), an endemic honeyeater, is also confined to one island, Little Barrier

Island. The species with the most tenuous hold on survival is the Chatham Island Black Robin (*Petroica traversi*), reduced to only seven birds in 1975, which were all transferred to another island where habitat was more suitable (Flack 1977). It is slowly increasing, and there are now 70 (Anon 1988).

Despite the depressing picture of declines and extinctions over the last 200 years there is reason for some optimism. There is now widespread public concern for New Zealand's native birds, legal protection for all and an extensive system of National Parks. Finally, the New Zealand Wildlife Service has carried out research on rare species and made dccisions, sometimes courageous ones, to manage dwindling populations. Small offshore islands, where the habitat is often unchanged and where introduced predators are absent, provide the key to the survival of many species. On less pristine islands habitat has sometimes been modified or predators and herbivores removed. There is always the risk that rats may reach these small refuges so that where possible healthy populations need to be established on several islands. Transfers between islands have had very positive results in the case of Saddlebacks, and appear to have been successful in the case of the Black Robin. An extensive captive-breeding programme of the Brown Teal (*Anas aucklandica*) is also having marked success.

On the other hand conservation of the Takahe is an uphill battle against red deer, stoats and poor quality food. Captive breeding is also being carried out. Probably the Kakapo will provc to be the hardest species to save, but then with ingenuity and luck, this too can be achieved.

Endangered and Extinct Birds on the Pacific Islands

Generally in the Pacific the more remote the islands, the fewer the species of landbirds there, the greater the proportion that are found only there and the more they have suffered from human settlement.

The Hawaiian Islands are very old, volcanic and very remote. They possess the most remarkable case of adaptive radiation of birds in an archipelago, the honeycreepers (Drepanididae) (Amadon 1950), which surpasses even that of Darwin's finches on the Galapagos Islands. A single species of finch, from Asia, diversified into at least 29 species including forms taking seeds, nuts, fruits, nectar and insects. Morphologically, the honeycreepers ranged from heavy-beaked finch-like forms through slender-beaked insectivores to nectar-feeders with long and sometimes curved beaks. The most remarkable species had a stout lower mandible and a long upper mandible which could be moved sideways. However, in terms of conservation, Hawaii is a disaster area with 68 of the endemic species or subspecies now extinct or on the brink of disappearing, including over half of the honeycreepers (Scott and Kepler 1985). Polynesian settlers, 1,000 years before Europeans arrived, also exterminated numerous species (Olson and James 1982). The familiar reasons of habitat destruction and introduced predators and herbivores are important but the single major factor is at present the Night Mosquito (*Culex quinquefasciatus*), which is a carrier of avian malaria. This disease now virtually restricts native birds to above 1,500 m, where mosquitos are rare. Fortunately, there is some evidence that honeycreepers are starting to adapt behaviourally and immunologically to reduce their susceptibility to malaria (van Riper *et al.* 1986).

The rest of Polynesia has suffered a similar fate, though most islands only had a few endemic species by the time of European settlement. The successful colonists of distant islands include rather surprisingly, rails and parrots. Rails have reached almost all islands, sometimes on more than one occasion and frequently became flightless, making them particularly vulnerable to hunting and to introduced predators. Parrots of the genus *Cyanorhamphus* have spread to Norfolk and Lord Howe Islands in the west to Antipodes and Macquarie Islands in the south and Tahiti in the east.

Norfolk Island provides yet another example of loss of endemic species from a small island (Schodde *et al.* 1983). Two species and three subspecies are believed extinct and the endemic subspecies of owl appears to be down to a single individual (Olsen 1986). A captive-breeding programme on the Lord Howe Island Wood-hen has increased its numbers over the last decade, though even now very little is known about the birds that have been released (Miller and Mullette 1985).

The birds of Fiji and the islands around New Guinea have fared better so far. New Guinea itself appears to have lost no species of bird yet and provides an opportunity to conserve a rich and unique tropical island avifauna in its complete form. Schodde (1979) listed 182 rare species or subspecies from the mainland of New Guinea. The greatest proportion of species is from isolated mountain ranges, where many species are naturally scarce or restricted. He listed a further 102 species from the Solomon Islands and many others from the smaller island groups. The most vulnerable species include several bowerbirds and birds of paradise. It is clearly of the highest international priority to ensure that the familiar pattern of habitat destruction, introductions and excessive hunting is not repeated in this region. The Papua New Guinean and Indonesian governments should be given tactful encouragement and support in their efforts towards conservation. Christmas Island, in the Indian Ocean, has so far retained an intact avifauna despite phosphate mining (Stokes 1988). Mining ceased in late 1987 and most of the rainforest on the island should shortly become national park. Christmas Island may escape the extinctions suffered by Lord Howe and Norfolk Islands.

Summary

On the whole, Australia's birds have not suffered badly from two centuries of European settlement, at least compared with its native mammals. However, there is a long list of rare or threatened species (Table 45) and there is an urgent need to find out more about these birds. Many may be naturally rare, as they have specific habitat requirements, whereas others may be declining due to human activities. The dedication of many national parks in recent decades is encouraging but some habitats, and the plants and animals in them, require active management to maintain the highest diversity.

New Zealand, and other Pacific islands, in contrast, have suffered badly from humans and the mammals they introduced. Many bird species have been lost, but increased awareness and ambitious research and management, give rise to hope that those that remain will survive.

11. Habitat Destruction and Modification in Australia

In most continental regions endangerment and extinction are the final result of the continual depletion of the native flora and fauna. If one can identify species that are particularly vulnerable to habitat loss or change while large areas of habitat remain, it may be possible to prevent them becoming endangered. Modification of natural habitat for economic reasons takes a variety of forms. Cropping requires total removal of native vegetation, grazing leads to loss of the shrub layer, major changes to the herb layer and thinning and lack of regeneration of the tree layer. Logging, clearing for wood-chips and mining may lead to short term removal of the vegetation followed by a period of at least partial recovery. Selective logging and use for military training or recreation may lead to local disturbance while still retaining the basic structure of the habitat.

In this chapter I discuss the effects of agriculture and clearing on bird assemblages, chiefly in woodland and forest, in an attempt to identify the sorts of species that suffer most, even though at this stage they are not considered to be endangered. Next, I shall discuss species that have benefited from human activity in Australia and those that are deemed to have become agricultural pests. Finally, I shall look at new habitats like pine forests and suburbs and discuss introduced species.

Removal of vegetation, typically either eucalypt woodland or mallee, to grow crops effectively removes nearly the whole avifauna. A very few species that lived in clearings in the original habitat may survive. There are economic reasons for retaining belts of vegetation even in the most intensive grain-growing areas, chiefly to reduce wind and water erosion and prevent salting. Unfortunately this has not always been done. Grazing by cattle or sheep inevitably results in some modification of habitat and where it is carried out intensively may lead to a major change in habitat. The ideal from the land-owner's viewpoint is usually grassland with scattered tall trees to provide shade and shelter. Small patches of woodland on poorer soils can also provide additional shelter and a supply of wood. Such patches can provide habitat for the original bird species and indeed are the only refuges over large areas of eastern and southern Australia (Saunders *et al.* 1987). Even where there are substantial state forests, national parks and reservoir catchments, the natural habitat has been fragmented.

Conservationists have likened this pattern of habitat fragmentation on continents to the situation on islands (Diamond 1975b; Wilson and Willis 1975). The smaller the island the fewer species that can survive. There is a clear relationship between number of species of birds on a patch of woodland or forest and patch size in Australia, just as has been found on other continents (Figure 39; Howe *et al.* 1981; Kitchener *et al.* 1982; Kitchener 1982; Howe 1984; Loyn 1985). One small patch of habitat is clearly not as good as a large area of original forest. Certain species do very poorly because they are large and require large territories, have specific habitat requirements or are predators at the top of the food chain. In most cases what remains is a patchwork of natural habitat in a sea of alienated land.

There is considerable dispute about the relative value for conservation of one continuous patch of habitat compared with a similar area consisting of many patches. Diamond (1975b) and Willis (1984) claim that single large areas are preferable to fragmented patches of the same total area. Higgs (1981) and Simberloff and Abele (1982, 1984) on the other hand maintain that this need not be so. In fact, which is better depends on the aims of conservationists. A series of small patches containing a variety of habitats may be as rich, if not richer, in number of species as one larger patch of the same area. Loyn (1985) found in Gippsland, Victoria that there were more species of open country and as many forest species in fragmented forest as in continuous forest of the same area.

To maximize the total number of species or the density of birds may not be the desired aim, however. The priority in conservation is usually to maintain the rarer species. By studying small patches of woodland surrounded by farmland and small plots within extensive woodland, Howe (1984) was able to show that several species appear to have an aversion to small isolated woods. Loyn (1985) studied a wider range of sizes of woodland and found that many species increased in frequency above about 10 ha.

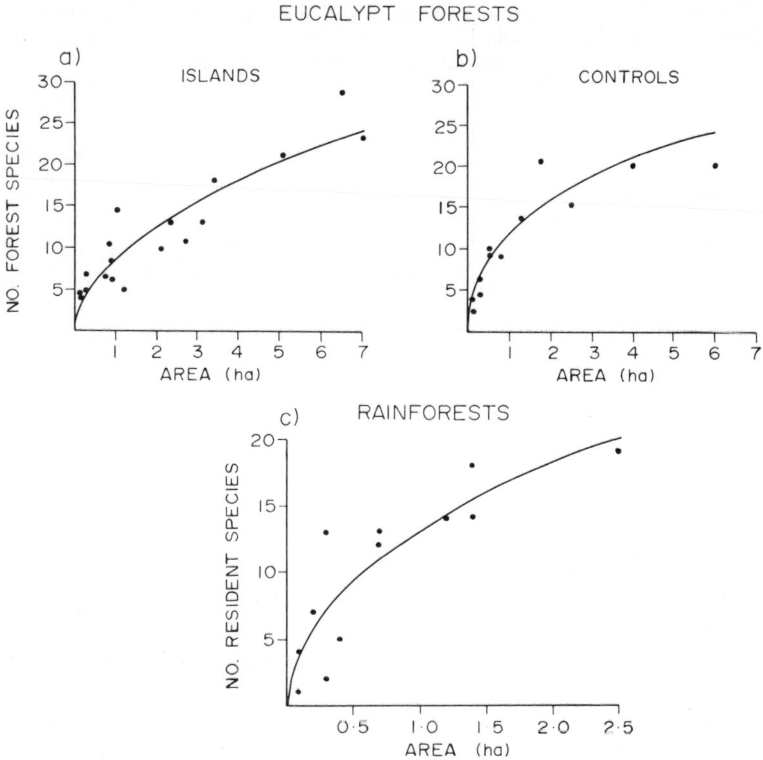

Fig. 39. Numbers of bird species in (a) "islands" of eucalypt forest, (b) control patches within continuous eucalypt forest, and (c) rainforest "islands" in northeastern New South Wales (from Howe *et al.* 1981; Howe 1984).

Kitchener (1982) separated species of birds, mammals and reptiles in nature reserves in the Western Australian wheatbelt into those that could cope with disturbance and fragmentation and those that could not. Examples of species that are relatively common in continuous woodland but rare in small fragments are given in Table 46. In most cases it is unclear whether these species simply dislike crossing open areas or whether there have been subtle changes to the habitat with fragmentation, such as exposure to drying winds or loss of understorey. Birds like honeyeaters also move between different habitats within a larger woodland.

Some species prefer the small woods to large areas of woodland (e.g., Eastern Rosella *Platycercus eximius*, Willie Wagtail *Rhipidura leucophrys*, Magpie *Gymnorhina tibicen*). Noisy Miners

Table 46. Species of birds that are found in continuous eucalypt forest and large fragments but less commonly on small fragmented patches (Howe 1984; Loyn 1985), and their status on four large isolated forested regions.

	Status in:			
	Tasmania	Kangaroo Island	Mt Lofty Ranges	South-west W.A.
Australian King-Parrot *Alisterus scapularis*	Absent	Absent	Absent	Absent
Sacred Kingfisher *Halcyon sancta*	Absent	Absent	*Present*	*Present*
Cicadabird *Coracina tenuirostris*	Absent	Absent	Vagrant	Absent
Eastern Yellow Robin *Eopsaltria australis*	Absent	Absent	Absent	Absent*
Satin Flycatcher *Myiagra cyanoleuca*	*Present*	Absent	Vagrant	Absent
White-browed Scrub-wren *Sericornis frontalis*	*Present*	*Present*	*Present*	*Present*
Red-browed Treecreeper *Climacteris erythrops*	Absent	Absent	Absent	Absent
Yellow-faced Honeyeater *Lichenostomus chrysops*	Absent	Absent	*Present*	Absent
White-naped Honeyeater *Melithreptus lunatus*	Absent*	Absent	*Present*	*Present*
Spotted Pardalote *Pardalotus punctatus*	*Present*	Vagrant*	*Present*	*Present*
Olive-backed Oriole *Oriolus sagittatus*	Absent	Absent	Vagrant	Absent

*Sibling species present.

(*Manorina melanocephala*) tend to frequent disturbed areas and may aggressively exclude typical woodland birds (Loyn 1987). In addition predation on nests may be more frequent in fragmented habitats, as species such as corvids are common in open country (Andrén *et al.* 1985) and wander into the edge of forests in search of food.

When eucalypt woodland is partially cleared and fragmented it tends to deteriorate. Loyn (1987) found that both the intensity of grazing and the time since fragmentation had a negative effect on the number of species in patches of woodland. In the tablelands of New South Wales eucalypt dieback is extensive (Ford 1985d). Dieback is an imprecise term meaning the loss of vigour of trees, usually with defoliation followed by resprouting from the trunk or branches. Successive bouts of defoliation can lead to death of the trees. Insects, particularly scarabs (Christmas beetles), leaf-beetles, caterpillars and sawfly larvae, are the major defoliators and most conspicuous cause of death. However, dieback is a complex process which probably started with the partial clearing of the country and grazing with resultant loss of understorey and changes to the herb layer and soil. (See Heatwole and Lowman 1986 for a popular account of dieback and Wylie and Landsberg 1987 for an excellent summary.) The resistance of the trees could have been weakened by changes to the soil nutrients and organisms, and the natural predators of herbivorous insects could have declined. There is some evidence that insectivorous birds play a part in the control of herbivorous insect populations (see Chapter 4). A decline of insectivorous birds could lead to more frequent outbreaks of the defoliators, all of which are native species, and some of which at least occasionally break out in natural undisturbed forest.

In most cases, then, extreme fragmentation of woodland or forest is a temporary state, which leads inevitably to the total loss of that habitat. There is some evidence that larger patches of woodland are more resilient to dieback but it is clear that landowners need to manage woodland on their property actively. The best ways to do this are only just being discovered. The numbers of individuals and species of birds decline dramatically with clearing and dieback though the relative importance of the two factors has not been defined (Table 47). Dieback is by no means confined to New South Wales.

Table 47. Bird communities on four sites suffering to varying degrees from dieback and clearing (Ford and Bell 1982).

Vegetation	Number of species	Density
Healthy woodland	47	22.5/ha
Moderate dieback	35	10.5/ha
Severe dieback	8	2.4/ha
Open grassland	6	0.8/ha

Saunders and Ingram (1987) have shown how clearing and dieback have together had a detrimental effect on the Carnaby's Cockatoo (*Calyptorhynchus funereus latirostris*) in Western Australia. These birds feed only on native shrub seeds and cannot breed successfully if their habitat is too fragmented, as food sources are too dispersed. Provision of corridors between patches of habitat that are visible from each other would help the species.

The theory of island biogeography states that the number of species on an island declines with declining area due to reduced dispersal and increased chance of extinction. Small populations are prone to extinction through chance fluctuations and genetic deterioration. Biogeographers believe that progressive decline of diversity will apply to national parks, most of which retain only a fraction of the original habitat. As noted earlier, Loyn (1987) found that the more years that a fragment had been isolated the fewer individuals and species of forest birds it held. That even very large refuges retain only a proportion of the species of the original habitat can be seen by looking at natural isolates. Tasmania, Kangaroo Island, the Mount Lofty Ranges and south-western Australia have all been part of the large belt of eucalypt forest and woodland in eastern and southern Australia. They have far fewer forest species than occur in forests in New South Wales or Victoria (Abbott 1973, 1974). It is interesting that many of the species that Howe and Loyn identified as poor dispersers to or survivors on small patches in New South Wales and Victoria do not occur on these large isolated areas (Table 46). Abbott (1978) studied birds on islands off the coast of Western Australia. He found that the islands were highly impoverished, which he attributed to the low immigration rate of most species. He warned that birds lost from isolated nature reserves will be lost forever.

The conservation parks of the Mount Lofty Ranges provide isolated patches of forest and woodland intermediate in size between Howe's small patches and the large islands. The number of species in them agrees well with the general pattern from islands of all sizes. Ford and Howe (1980) used this species-area relationship to predict that the current reserves in the Mount Lofty Ranges were totally inadequate to conserve the full forest avifauna that they presently contain. Species that were already scarce and well isolated from the original populations were suggested to be particularly at risk. Even though the application of the theory of island biogeography to conservation has been severely criticized it does provide some answers on how to reduce the rate at which species are lost. MacArthur and Wilson (1967) envisaged that an equilibrium number of species on an island is reached when immigration and extinction rates are equal. Active conservation should increase immigration rates by providing corridors and if necessary reintroducing species. This would also maintain genetic variability. Local extinction rates can be reduced by habitat modification to maximize population sizes and buffer fluctuations between years. The populations of each of the species at risk would need to be monitored and their habitat preferences accurately measured. In some cases it may be necessary to reduce or remove predators or competing species from reserves. Bell Miners (*Manorina melanophrys*) may exclude the endangered Helmeted Honeyeater (*Lichenostomus melanops cassidix*) from rich patches of food in Victoria (Wykes 1985) and Striated Pardalotes (*Pardalotus striatus*), when common, can have a detrimental effect on the Fofty-spotted Pardalote (*P. quadrigintus*) in Tasmania (Woinarski and Bulman 1985).

What has happened in the Mount Lofty Ranges is similar to what is happening throughout eastern Australia. The important difference is that the Mount Lofty Ranges originally occupied a smaller area and habitat loss has proceeded further. What happens to the avifauna there over the next few decades and the success of measures taken to counteract any further loss of species will provide invaluable knowledge for the conservation of birds throughout the forested areas of Australia.

The pattern of habitat loss and fragmentation that I have discussed at length for eucalypt forests and woodlands is occurring in all major habitats in Australia and indeed throughout the world. Kitchener *et al*. (1982) found that the area of patches of mallee in Western Australia closely predicted the number of bird species. Consequently even though many large and reasonably pristine national parks have been declared in mallee country these will not guarantee the survival of the rich avifauna of this habitat. Rainforest is a scarce habitat anyway in Australia and a large proportion of it has already been cleared or logged. Again small patches are inferior to larger areas in terms of bird species (Howe *et al*. 1981).

Although this picture of the inadequacy of national parks and nature reserves to conserve the total flora and fauna of a region is a depressing one, I believe that one does not have to be pessimistic. First, the wetter habitats in particular have been greatly fragmented in the past during the arid periods of the ice ages. Species have perhaps been lost but many, probably most, survived. What is more, some have diverged in their isolated refuges. Fragmentation may increase extinction rates but it can also increase rates of speciation. In fact one view of speciation is that it typically takes place in small isolated populations which undergo genetic revolutions (Templeton 1980). Second, birds and other organisms can adapt to change, providing that it does not take place too rapidly. Evidence for this can be seen in the British Isles where barely a scrap of natural habitat remains, but where birds are still common and indeed the number of breeding species has increased over the last century. I shall discuss how birds are coping with new habitats later in this chapter. Finally, active conservation can prevent local loss of species. This will require long-term population studies, which I have already stated to be a high priority in Australia. These could easily be carried out by amateur ornithologists. Habitat modification, should it be necessary, is perhaps best carried out by wildlife authorities. Saunders *et al.* (1987) argue strongly for a concerted effort into research and management of fragments. In particular, they stress the importance of communication between researchers and managers and between both and local landowners.

There is no reason why increasing human populations and use of the environment has to lead to loss of bird species. Järvinen and Ulfstrand (1980) discussed the changes over 120 years to the bird fauna of northern Europe. Each country showed a net gain of 2.2 species per decade, despite this being a period of human population increase and unprecedented economic expansion. The main reasons for gains, which have accelerated in recent decades, have been protection from persecution and an increase in the heterogeneity of the habitat. The latter includes partial clearing of extensive forests for agriculture, planting of trees or regrowth of scrub in areas previously grazed and the eutrophication of some wetlands due to the indirect effects of agriculture. Species have been lost though. These were typically rare at the beginning of the period, were mostly non-passerines and large. Many, such as the raptors, were persecuted but others suffered from habitat change, especially loss of old forest and intensification of agriculture. Most of the lost species fed on vertebrates or large invertebrates. Even though this region is only a part of a larger one, the Palaearctic, the results are valuable in identifying species that are likely to be at risk in continents as a whole. Furthermore, Scandinavia still possesses large areas of natural habitat, albeit heavily exploited and managed ones. There is also a generally enlightened view to conservation.

The Effect of Logging and Mining on Birds in Forests

Selective logging involves the removal of large usually healthy trees for timber. It is less detrimental in the short term than clear-felling to wildlife, though mature trees are particularly important for some species. An increasing tendency is to exploit eucalypt forests by removing all trees and converting the poorer quality wood into wood-chips. In the most extreme cases only bare earth or litter remains. Clearly the avifauna is greatly affected by this procedure. Several studies have looked at the effect of logging, including wood-chipping on the bird communities of eucalypt forests in southeastern Australia (Loyn 1980; P. Smith 1985a, b; Kavanagh *et al.* 1985; Recher *et al.* 1987a). Over the first four years the habitat changes from being very open to a dense low scrub dominated by acacia. The birds change accordingly. Recently logged areas have about half the number of individuals and species of unlogged areas. Initially, birds like Swallows (*Hirundo neoxena*), Pipits (*Anthus novaeseelandiae*), Willie Wagtails and Jacky Winters (*Microeca leucophaea*) are common as these species forage in open areas. As the shrub layer develops these species decline, Chestnut-tailed Heathwrens (*Hylacola pyrrhopygia*) colonize and Superb Fairy-wrens (*Malurus cyaneus*) and Yellow-faced Honeyeaters (*Lichenostomus chrysops*) increase. By 9–17 years after logging a tall shrub layer has developed with emerging eucalypts. The bird abundance and community is similar to that of unlogged forest. Birds like Lyrebirds (*Menura novaehollandiae*) and Eastern Whipbirds (*Psophodes olivaceus*), which like dense habitat, are common, but inhabitants of the eucalypt canopy such as the Gang Gang Cockatoo (*Callocephalon fimbriatum*), Rufous Whistler (*Pachycephala rufiventris*) and the pardalotes are still less common than in unlogged forest. Forest on ridges takes longer to recover than that in moist gullies. Two other groups may take a long time to return fully after logging. The robins (*Petroica, Eopsaltria*) require a rather open ground layer and trees for their sit-and-pounce technique of feeding. Young forests are too dense for them. Second, hole-nesters require mature trees which are most likely to have developed safe, dry holes. The cockatoos, owls and Red-browed Treecreeper (*Climacteris erythrops*) are particularly affected.

Peter Smith (1985a) made two proposals to increase the ability of birds to recover from logging. First, he suggested that a few large trees should be left in areas that are clear-felled. These will provide some nest holes, as well as substrate for canopy feeders. They are also likely to have mistletoes which provide nectar for honeyeaters and fruit for Mistletoebirds (*Dicaeum hirundinaceum*). In addition they provide perches for pouncing and sallying flycatchers. These perches could also encourage raptors. In Swedish forests Buzzards (*Buteo buteo*) prefer clear-felled areas with perches and may reduce outbreaks of voles which can damage the forest. Thus, leaving a few mature trees in logged areas need not be expensive and could possibly be economically valuable too. It needs to be determined how many trees are required and where they are situated. Smith's second suggestion was to retain rich patches of gully forest. These are prime breeding areas and can provide refuges from fire. They also tend to contain a wide variety of plants including many that produce fleshy fruits. These are fed upon by bowerbirds, pigeons and others, which may be scarce in the ridge forests.

Thus, if some allowance is made for wildlife there is no reason why eucalypt forests that are used intensively for logging and wood-chipping need not retain a diverse avifauna. It is important that large areas of unlogged forest are retained for those species that take a long time to re-appear after logging and to act as a source for reinvasion of logged areas as they recover. Recher *et al.* (1987a) suggested that by logging alternate coupes on a long rotation and linking all areas to large reserves, the effects of logging on wildlife can be minimized. There is still a great need for further research on the bird assemblages of logged forests right up to the time when the second episode of logging is contemplated. The details of wildlife management in areas subjected to forestry still need to be worked out.

Mining has a more severe effect on the environment as not only is the vegetation removed but so too is the soil and underlying rocks. Collins *et al.* (1985) studied the recolonization of Jarrah forests in Western Australia that were mined for bauxite. The bauxite is only removed to a depth of about 5 m. Originally, after mining the topsoil and overburden were replaced and some trees planted, often species that were not endemic to the area. The steep sides to the mined areas, of compacted clay, and poor viability of seeds in the topsoil led to rather poor regeneration of the vegetation. Now, the pit walls are reduced, fresh topsoil is added from an area that is about to be mined and soil and clay are ripped. Fresh seeds are also added and trees planted. This has resulted in much more rapid revegetation. Recovery of the bird community is similar to that after logging. Species that prefer dense understorey, like White-browed Scrub-wrens (*Sericornis frontalis*) and Red-winged Wrens (*Malurus elegans*), increased rapidly. On the other hand the ground-feeding robins and bark-feeders were uncommon. Because of the nature of mining, mature trees cannot be left so that hole-nesters are likely to take a very long time to return unless nest boxes are placed in the regenerating forests. Unmined areas nearby are needed to ensure that the full range of bird species survive in the area.

Although mining is at first devastating to the forest, recovery of the vegetation can be rather rapid using modern techniques. With further research these could be refined. For instance a range of non-indigenous eucalypts and shrubs have been planted first and are thriving. Their effect on the fauna still needs investigating. Dieback, caused by the fungus *Phytophthora cinnamomi*, is a constant threat where there is disturbance to the soil, though species that are currently resistant are used. Nectar-bearing shrubs could also be planted, although some are susceptible to *Phytophthora*, and some provision could be made for hole-nesters.

Birds that have Benefited from Human Activities

By no means all birds have suffered since Europeans arrived. Although the parrots include many declining, even endangered, species, some species have become very successful (Smith 1979b). The Galah in particular has expanded dramatically following agricultural development (Saunders *et al.* 1985). Originally Galahs occurred in the semi-arid woodlands of central Australia wherever there were trees large enough to have nest-holes. They fed on grass and wattle seeds on the ground, but were also dependent on water. Clearing of woodland and forest has provided suitable habitat which in many areas has allowed Galahs to spread to the coast. In grain-producing areas they have an abundant source of food for most of the year and farm dams have greatly increased the availability of water. Galahs also have a rather high reproductive success for such a large bird and adults can live for a long time. They have a social organization similar to that of the Eastern Rosella (see Chapter 8). There is a non-breeding flock of young birds and it is these birds that progressively colonize new areas (Rowley 1983).

Other parrots that have benefited from agriculture include the Corellas, Cockatiel and some of the rosellas. In some areas they are now regarded as pests of cereal and oil-seed crops (Table 48). Damage to crops is mostly caused by the non-breeding flocks though sometimes breeding birds will join flocks. Various attempts to control parrots and cockatoos have been used in the past, chiefly shooting and poisoning. These have mostly removed only a small proportion of non-breeding birds which has had a negligible effect on the population. Where breeding birds join

Table 48. Some parrots that may damage crops.

Species	Crop eaten	Reference
Baudin's Black-Cockatoo *(Calyptorhynchus baudinii)*	Apples	Saunders *et al.* 1985
Galah *(Cacatua roseicapilla)*	Wheat, sunflower etc.	Forshaw 1981
Long-billed Corella *(C. tenuirostris)*	Wheat, maize etc.	Emison *et al.* 1981
Little Corella *(C. sanguinea)*	Sorghum	Beeton 1977
Sulphur-crested Cockatoo *(C. galerita)*	Sunflower, sorghum, wheat	Forshaw 1981
Cockatiel *(Nymphicus hollandicus)*	Sorghum, sunflower, wheat	Jones 1987
Rosellas *(Platycercus)*	Fruit, grain	Breckwoldt 1983
Ringnecks *(Barnardius)*	Fruit, grain	Breckwoldt 1983

flocks and most of the population joins a few large flocks, as occurs in some Black-Cockatoos, mass shooting and especially poisoning could have a devastating effect on the population (Saunders *et al.* 1985). The White-tailed Black-Cockatoo occurs in two forms, a short-billed form (now called a subspecies of the Funereal Black-Cockatoo or Carnaby's Cockatoo and a long-billed form (*C. baudinii*) (Saunders 1979). The former may be a nuisance in pine nurseries, whereas the latter occasionally damages fruit crops.

The result of more detailed studies on parrots and cockatoos that are deemed pests has been two-fold (references in Table 48). For a start the problem caused by the birds has often been over-estimated, in monetary terms it is not great for any crop as a whole (Beeton 1975). In times of world-wide grain surplus surely a small share taken by birds can be tolerated by the country. However, individual farmers may be particularly badly hit. With careful planning and neighbourhood co-operation, damage can be minimized (Breckwoldt 1983). Typically, small crops in areas where agriculture is diverse may be badly damaged. Such crops are particularly vulnerable if there are trees nearby or even in the crop, and where growth has been patchy. Early or late crops are more likely to be attacked than those maturing in the middle of the season. Reducing, or at least sharing, the damage caused by birds to a crop can be achieved by co-ordination and synchronization of planting over a wide area. Flocks could also be deflected to decoy crops which ripen just before the main harvest, and are planted to be particularly attractive (Beeton 1975). As these are depleted, birds could be left in peace to feed on stubble, where seeds are plentiful and conditions good for detecting predators. Obviously details of management need to be carefully worked out with a certain amount of trial and error, but it should be possible in most cases to minimize or spread the damage to main crops.

Several other species have increased their ranges and numbers after clearing of forest and woodland. These include Crested Pigeons (*Ocyphaps lophotes*), Australian Magpies and Magpie-Larks (*Grallina cyanoleuca*). Other species have increased in the inland with provision of water in otherwise dry areas. The seed-eating parrots, pigeons and finches have done best, but Emus (*Dromaius novaehollandiae*) have probably also increased as may have some honeyeaters where the habitat has not been greatly altered otherwise.

Ducks and geese are sometimes a problem in rice fields; eagles, crows and ravens may kill lambs, and parrots, silvereyes and starlings consume grapes and other fruit (Breckwoldt 1983). Many of these birds can also be of value to land-owners; for instance ducks can be hunted and eagles consume rabbits, and, again, careful management can reduce damage.

New Habitats

During the last century two other new habitats have also increasingly become available for birds. These are pine-forests and urban areas. Well managed pine-forests (mostly *Pinus radiata* but also often *P. carribea*) contain approximately half the number of species and half the number of individuals found in native forest (Driscoll 1977; Friend 1982; Gepp 1986). Birds occurring in pine-forests fall into two groups. Probably the majority depend on remnants of the original habitat (e.g., Striated Thornbills *Acanthiza lineata* – Recher, Davies and Holmes 1987); they feed on native vegetation along tracks and firebreaks, where pines have not grown well or where it abuts the pine-forest. Mature pine-forests may have a well developed understorey of wattles and sapling eucalypts which some birds favour. Nectar-feeders in particular require native vegetation. Some species on the other hand appear to be able to cope with areas purely of pines (e.g., Brown Thornbills *Acanthiza pusilla* – Recher, Davis and Holmes 1987).

Many of these are ground-feeders like White-winged Choughs (*Corcorax melanorhamphos*) and Yellow-rumped Thornbills (*Acanthiza chrysorrhoa*), but flycatchers and robins are successful where older pines are well-spaced. Management of pine-forests should include small patches of natural habitat and some natural regeneration should be tolerated. This would encourage birds and other wildlife, which is one of the stated aims of the forester. In addition, it is possible that a healthy population of insectivorous birds and mammals may reduce the likelihood of outbreaks of insect pests on the pines. More work is needed on this aspect of the management of pine plantations.

Australia's cities have sprawled giving a low human population density with many parks and well-vegetated gardens. The trend over the last twenty years has been to grow native trees and shrubs rather than exotic conifers and deciduous trees. The result has been a reasonably diverse avifauna. One of the most detailed studies of birds in cities was that of Green (1984, 1986) in Melbourne. She found that the number of species and density of native birds increased as one went from the city centre through old suburbs to new ones, whereas diversity and density of exotic species declined. The large buildings and often deciduous trees or man-made perches in the centre are the preserve of introduced species like House Sparrows (*Passer domesticus*), Starlings (*Sturnus vulgaris*) and Common Mynahs (*Acridotheres tristis*). As the amount and variety of native vegetation increases so does the proportion and overall number of native bird species. Many suburbs resemble open woodland with well-spaced eucalypts and extensive grass. Birds common there include Red Wattlebird (*Anthochaera carunculata*), White-plumed Honeyeater (*Lichenostomus penicillatus*) and Magpie-Lark, which are typically at home in savannah woodland. Where many native shrubs have been planted small honeyeaters like Eastern Spinebills (*Acanthorhynchus tenuirostris*) and Superb Fairy-wrens thrive. Lorikeets can be abundant among flowering eucalypts. Suburbs contain hazards that are rarely experienced in more natural habitats, like traffic, cats and windows. These take a steady toll of birds but behavioural flexibility or even genetic change can lead to adaptation to cope with these. Magpie-Larks seem to be particularly adept at foraging at the edge of busy roads. Some species (pardalotes, cuckoos and kingfishers) seem to be particularly prone to colliding with windows, whereas others are able, or have learned, to avoid them.

Daryl Jones studied two rather different urban areas, an inland city, Wagga Wagga (Jones 1981), and a coastal tropical city, Townsville (Jones 1983). He found thriving bird communities in both. Again, the amount of native vegetation strongly influenced the diversity and types of birds present. Bare new suburbs had few birds but as these developed along the succession through shrubland to woodland the density and range of birds increased. A "mature" suburb in Townsville had a similar diversity of species and greater density of individual birds than natural forest nearby, and even suburbs 10 years old were as diverse as the original habitat. Introduced species made up about half of the individuals though. These were mostly ground-feeders and seed-eaters, so native species feeding in this way may be excluded.

Some native species do well in suburbs whereas others often shun them or cannot survive there. Successful species are often medium-sized, such as the larger honeyeaters, parrots and Figbirds *(Sphecotheres viridis),* whereas small species are poorly represented. Miners can become common and will drive out many other species, whereas butcherbirds (*Cracticus*) and Kookaburras (*Dacelo*) prey on adults and nests. Smaller birds are more successful where dense understorey is provided or a full canopy has been retained. Species that do not do well in the suburbs are often the species that are most affected by logging or which tend to occur in extensive forest but not in fragments (Table 45). On a national scale these are the species that are most dependent on large, relatively undisturbed national parks.

One of the most interesting studies of urban birds is that co-ordinated by Bell (1986c) in the developing towns of Papua New Guinea. Most of these towns are small and very recent (post World War II). They are mostly well-endowed with native and exotic vegetation, which is fast growing. All towns have a rich avifauna, which is highly dynamic. Most common species come from open habitats or forest edge rather than rainforest. New species are colonizing and those already established are changing their habits. Honeyeaters and lorikeets have been very successful as there are many flowering trees, though frugivores have done less well. Bowerbirds and birds of paradise occur in towns and could become common if persecution was reduced and fruiting trees planted. Another feature of the urban avifauna in Papua New Guinea is the number of wintering migrants from Australia and to a much lesser extent from Asia. There is an almost total lack of exotic species and it is to be hoped that any proposed introductions are strongly resisted.

The important feature of this study is that it represents the early stages of colonization of new habitats. Repeated studies in the future will be of immense value in helping to understand how bird assemblages are constructed and how rapidly species can change behaviourally and genetically to exploit new resources.

Introduced Birds

In common with most parts of the world there have been numerous attempts to introduce new species of birds to Australia. Long (1981) listed a total of 71 exotic species that were released into Australia, of which 25 have probably become established. Fifteen species are given maps in the RAOU Atlas (Blakers *et al.* 1984), whereas a remaining 10 occur only as small isolated populations. They are all listed in Table 49, with approximate dates of introduction and for the more successful species the area of the continent

Table 49. Introduced species that have become successfully established in Australia. The proportion of 1° squares occupied is given as a measure of success (from Blakers *et al.* 1984).

Species	Origin	Date introduced	Success
Mallard (*Anas platyrhynchos*)	England	1860s	10%
Feral Pigeon (*Columba livia*)	Europe	before 1873	37%
Spotted Turtle-dove (*Streptopelia chinensis*)	China/S.E. Asia	1860s	11%
Laughing Turtle-dove (*S. senegalensis*)	Africa/S. Asia	1898	4%
Skylark (*Alauda arvensis*)	Europe	1850s	11%
Red-whiskered Bulbul (*Pycnonotus jocosus*)	S.E. Asia	1880	1%
Blackbird (*Turdus merula*)	England	before 1862	14%
Song Thrush (*T. philomelos*)	England	1857	1%
Goldfinch (*Carduelis carduelis*)	Europe	1857?	15%
Greenfinch (*C. chloris*)	Europe	1863	7%
House Sparrow (*Passer domesticus*)	Europe	1863	41%

Table 49 continued

Species	Origin	Date introduced	Success
Tree Sparrow (*P. montanus*)	China	1863	2%
Nutmeg Mannikin (*Lonchura punctulata*)	S.E. Asia	c. 1930	5%
Common Starling (*Sturnus vulgaris*)	Europe	late 1850s	30%
Common Mynah (*Acridotheres tristis*)	S. Asia	1862	7%
Very small populations in feral state (< 1% of 1° squares occupied), some may be extinct.			
Ostrich (*Struthio camelus*)	Africa		
Mute Swan (*Cygnus olor*)	Eurasia		
Peafowl (*Pavo cristatus*)	India		
Feral Chicken (*Gallus gallus*)	S.E. Asia		
Common Pheasant (*Phasianus colchicus*)	W. Asia		
California Quail (*Lophortyx californicus*)	W. U.S.A.		
Chukar Partridge (*Alectoris graeca*)	S. Europe		
Black-headed Mannikin (*Lonchura malaca*)	S.E. Asia		
Red Bishop (*Euplectes orix*)	Africa		
White-winged Widowbird (*E. albonotatus*)	Africa		

occupied. The main period of releases was in the late 1850s and 1860s, when acclimatization societies imported and liberated many European birds particularly around Melbourne and Adelaide. The introductions were to make home-sick migrants feel more at home. Most species are small and either attractive or closely associated with man in Europe. Some introductions were game-birds to provide sport, but these mostly failed. One of the most unusual introductions was the Ostrich, farmed in South Australia from 1869 onwards for the feather trade. Birds were released after the First World War and a few hundred survive in the Flinders

Ranges. A few species colonized from escapees from captivity, the Spotted Turtle-dove and Laughing Turtle-dove escaping from Adelaide and Perth Zoos respectively. A few Australian birds have also been introduced to new parts of the country, for instance the Sulphur-crested Cockatoo and Laughing Kookaburra *(Dacelo novaeguineae)* to Western Australia and the Superb Lyrebird to Tasmania.

Almost without exception the introduced species occupy cities, suburbs or intensive farmland. The Starling breeds in tree-holes on the edge of woodlands and forests; probably only the Blackbird has penetrated well into natural habitats. It has spread north and west during the last 50 years (Blakers *et al.* 1984). There is little evidence that exotic birds have displaced native birds from native vegetation. However, they have most likely reduced the opportunities for native species to colonize new habitats, such as suburbs, or adapt to new habitats. Most of the more successful introductions are seed-eaters, though the thrushes and Bulbul combine invertebrates and fruit in their diets. This combination is an unusual one in at least eucalypt woodlands and forests, though bowerbirds and to some extent honeyeaters consume invertebrates and fruit. Both these native groups have been rather successful in suburbs and small towns, possibly more so where the Blackbird and Bulbul are absent.

Ornithologists have generally neglected exotic species, but studies on them would be valuable for two main reasons. One is their impact on human activities such as agriculture and on native species, and the second is that as species colonizing a new environment they show potential for rapid evolution. Natural selection is likely to operate differently in a new environment. In addition, as usually only a few individuals were introduced, they would only have contained a proportion of the genetic variability of the original population. The effects of chance would allow rapid changes to take place in a short period or alternatively the restricted gene pool could slow the rate of evolution. One hundred generations, about the time that many exotic birds have been in Australia, have proved to be enough for considerable changes in laboratory populations. House Sparrows in both North America (Johnston and Selander 1971) and New Zealand (Baker 1980), have changed morphologically from their European ancestors, and now vary geographically in their new countries. Similar changes could have taken place in most exotic species, as could have changes in ecology or behaviour, which may have some genetic basis.

Finally, studies on exotic species may allow them to be controlled or perhaps if desirable to be exterminated. Perhaps those species with very small populations should be eliminated before they reach unmanageable population sizes. Alternatively, they could be studied intensively as models for endangered species that have been reduced to small numbers.

New Zealand, and especially Hawaii, have experienced large numbers of introduced birds, though surprisingly the percent of attempted introductions that were successful is no more than in Australia (Long 1981). In many parts of New Zealand and Hawaii exotic birds far outnumber native ones. This replacement has apparently not occurred because the introduced species are competitively superior to native ones. Rather, endemic species seem to have been reduced by introduced predators and disease in combination with habitat modification. Exotic species have then been able to occupy not only cleared areas but also to move into native forests (Diamond and Veitch 1981).

Summary

There is a need, especially in Australia, to practice conservation at the level of the biotic community. Logging and mining should be carried out in such a way that all animal and plant species are able to return after human exploitation is over. Where agriculture requires the removal of native vegetation, representative patches should be retained or where necessary reconstructed (Breckwoldt 1983 discusses methods for doing this). There is a need for research into the sizes, shapes and configuration of such patches in order to maintain healthy and diverse communities.

There should be an endeavour to make new habitats, like pine-forests and suburbs, as suitable as possible for native birds. Any new introductions of birds should be strongly resisted and an effort made to eliminate exotic species that at present only have small populations before they become unmanageable. Research into conservation and management of the continent's birds is clearly of the highest priority in the final decade of the twentieth century.

12. Bird Ecology in Australia – A Synthesis

In this final chapter I summarize the contributions that studies in the Australian region have made to an understanding of bird ecology and to ecological theory as a whole. To what extent do they provide straight-forward support to current theories, and where do they lead to extension or modification of theory? Having discussed the stage that has been reached in various fields, I then suggest the questions that should be asked next.

Many of the broad patterns were developed in the 1950s by Keast and Frith, but it took the detailed studies of Richdale, Serventy, Rowley and Carrick to set the population and behavioural ecology of Australian birds on a sound course. The field of community ecology was to emerge in the late 1960s and 1970s, through the work of Recher and Kikkawa. The late 1970s and 1980s have seen a proliferation of methodical studies. The majority of these have been descriptive and quantitative, and it is only more recently that attempts have been made to perform experiments. In addition, applied problems, such as the conservation of endangered species, are now being tackled.

Distribution

A knowledge of the distribution of Australian birds has been gained in two main phases. Keast described and interpreted the ranges of most species in the 1950s (Keast 1961). More recently the RAOU Bird Atlas provided the details of distribution for all species, at least on a continental basis (Blakers *et al.* 1984). Keast's interpretations were evolutionary as well as ecological. Many species are restricted to particular habitats, which provide food and safe nesting sites. They may also be constrained by barriers. Barriers in the past have led to the isolation and differentiation of populations into subspecies, or over a longer period, into species. In contrast to other continents Australia has no mountain ranges that are effective as major barriers to the movements of birds.

Also, unlike the northern continents glaciation has rarely, if ever, provided a barrier. The major impediment to distribution has been the arid centre of the continent especially where this has reached the coastline. As this expanded and contracted during the fluctuating Pleistocene epoch, birds retreated to and spread out from peripheral refuges. In some cases species could have been lost from some refuges; in others, populations diverged sufficiently to become reproductively isolated when next they met. Many of the patterns of distribution at the level of the subspecies and species can be traced to this period. Expansion from and contraction into refuges in times of fluctuating rainfall were also important in Africa and even in South America where the great Amazonian forests have varied greatly in extent (Williams 1975). The overall picture is by no means unique to Australia.

Whereas Keast described the habitats of many species, and the match between vegetation and distribution can often be seen in the Atlas, few studies have really quantified a species' requirements. The genus *Meliphaga* (including *Lichenostomus*) provides a classic example of congeners that are separated by habitat. One of the most detailed studies of habitat preference is that of Ferrier (1985) on the rare Rufous Scrub-bird (*Atrichornis rufescens*). A precise description of the habitat requirements of rare species is obviously valuable for future management of such species. Ferrier's analysis could be used as a model for future studies of rare species. A different approach to habitat preferences is to examine the physiological requirements and tolerances of a species or population and here the studies of Williams and Main (1976) and Ambrose and Bradshaw (1988a,b) provide guidelines for future work.

Another major barrier both within and between continents is the sea, though birds are more able than many organisms to cross it. Both New Guinea and Tasmania, at times of low sea levels, were connected to Australia allowing the invasion of new species, mostly from north to south. At other times opening of the Torres and Bass Straits has allowed isolation and divergence. The numerous islands of the Pacific have almost invariably been continuously isolated from continental land masses. The numbers of bird species colonizing them declines with distance from the mainland. Isolated archipelagoes like New Zealand, and in particular Hawaii have, or had, depauperate avifaunas which included many unique forms. Islands in the Pacific have been of major importance in the development of the theory of island biogeography, in which the number of species reaches an equilibrium through a balance between immigration and extinction (MacArthur and Wilson 1967). Diamond (1975a) employed data on the distribution of birds on islands off New Guinea to show how some species are good dispersers to small islands, though absent from large islands, whereas others are restricted to faunistically rich islands. He has argued strongly that interspecific competition is the major force in preventing or restricting coexistence between similar species. Others (e.g., Simberloff 1978) believe that competition is not important. There is a need for more detailed work to illustrate direct competition for resources in coexisting species or at least depletion of resources where species do not overlap.

Community Ecology

For much of the history of birds the sea has provided a major barrier between Australia and other continents. A southern route of invasion was closed some 40 million years ago and a northern one has progressively opened over the last 10 million years. Perhaps surprisingly the most important groups of passerines arrived and diversified just before or during this period of extreme isolation. The Australian treecreepers, warblers, robins and wrens, as well as the honeyeaters, lyrebirds and bowerbirds evolved here, isolated from ecologically similar species elsewhere (Sibley and Ahlquist 1985).

The massive adaptive radiation that responded to the numerous opportunities in the forests and scrubs provides a chance to ask a wealth of ecological questions. Do different continents develop similar bird assemblages in similar habitats? Are there ecological equivalents among species? How many morphologically unique species are there? Do they occupy niches that are not present on other continents? Are some ecological types missing? Do species partition resources in the same way on different continents?

In this book I have emphasized landbirds because those in Australia are most different phylogenetically from those on other continents. Freshwater birds and seabirds are similar on all the world's continents, though all regions have some unique species. Eucalypt forests and woodlands are the most distinctively Australian habitats and their bird assemblages have been reasonably well studied. In them one finds foliage-gleaners, bark-foragers, flycatchers and species that pounce on prey on the ground as well as walk along the ground seeking food. The main foraging guilds that one finds in the Northern Hemisphere are represented. What is more,the species occupying them have often been given the same names as their northern counterparts, despite the fact that they are unrelated.

Thus, species that are ecological equivalents are to some extent morphological equivalents. This convergence should not be overemphasized though. None of the Australian forest birds has been subjected to the detailed examination relating structure to foraging behaviour that has been carried out in Europe (e.g., Norberg 1979). To take one guild in detail, among the bark foragers the Sittella (*Daphoenositta chrysoptera*) and climacterid treecreepers are to some degree equivalents to nuthatches (*Sitta*) and certhiid treecreepers. The Sittella does not eat seeds though. There are no true woodpeckers in Australia, nor species that have evolved to look and behave like woodpeckers. The treecreepers eat ants and behave rather like small woodpeckers, whereas at the other end of the scale Black-Cockatoos (*Calyptorhynchus*) chew through bark and wood to reach larvae. No birds drill into the wood. Is this because woodpeckers were unable to fly the 200 km to reach Australia? Were Australian birds unable to evolve the skeletal adaptations necessary to hammer into wood? Or are the opportunities for a "woodpecker" simply not available in eucalypt forests? I find none of these suggestions convincing at present. There does seem to be one uniquely Australian bark-foraging niche. Shrike-tits (*Falcunculus*) feed on peeling or shredding bark, which is abundant in eucalypt woodlands and forests. They have evolved powerful beaks to crush and strip the bark to reach insects. It is interesting that the early bird-watchers, in naming the species, were wrong with both its relationships (tit) and its foraging niche (shrike)!

Two other major food resources are available in eucalypt forests that are of minor importance in Northern Hemisphere forests. These are nectar and a range of other carbohydrates. The honeyeaters (Meliphagidae) have diversified and now monopolize these and are the dominant family in many Australian habitats, both in density and behaviour. The pardalotes *(Pardalotus),* and some thornbills (*Acanthiza*) are also dependent on the sugary secretions and shells of scale-insects and psyllids. The former have no obvious equivalent in Europe and North America. I have argued that the predominance of nectar and of bird-pollination in temperate Australia is the result of a series of historical and geographical accidents. The sclerophyllous flora developed in infertile patches of an old landscape that was beginning to dry out. Its patchy distribution necessitated a mobile pollinator that promoted outbreeding. The rather equable climate of the island continent permitted year-round growth and winter-flowering, despite the fact that, or perhaps because, insects were relatively inactive at this time. In the absence of large bees, small birds provided the most reliable service as pollinators. I hope that the reader will find this explanation unconvincing and will think of ways to test rigorously the various steps in the argument.

In contrast to the abundance of nectar, fruit is scarce in eucalypt forests. The honeyeaters sometimes eat it, as do fruit-doves (*Ptilinopus, Ducula*) and bowerbirds (Ptilinorhynchidae), which are more dependent on rainforest. The Mistletoebird (*Dicaeum hirundinaceum*) is the only true frugivore in eucalypt woodland and forest. It belongs to the flowerpeckers, an oriental group, and is probably a recent arrival in Australia. Why have so few of the ancient Australian assemblage started to eat fruit? Why has only one flowerpecker invaded Australia? Why has it not diversified? Milewski (1986) suggested that the low fertility of the soil in Australian sclerophyllous habitats has reduced the scope for producing large fleshy fruits. If plants require either mobile pollinators or mobile seed dispersers to ensure optimal spreading and mixing of genes, then the former option is less expensive when nutrients are scarce. The plants need to produce only a sugar solution rather than more plant tissue. Again this is a complex argument which future workers need to test, refine or reject.

Overall then bird assemblages of Australian forests and woodlands are similar to those of the Northern Hemisphere in some respects, while different in others. Some guilds are better represented and a few niches are unique, because the habitat provides unique opportunities. A few resources appear to be poorly exploited but it is not clear how readily available they are. Although comparisons between bird assemblages in Australia and elsewhere have been made descriptively there is a need for more quantitative comparisons using multivariate techniques.

Given that there are differences in the communities, do the same ecological rules apply in Australia and the Northern Hemisphere? Species partition resources among themselves on all continents, by differing in foraging substrate, method or food. Similar species replace each other geographically, in neighbouring habitats or in the case of the whistlers (*Pachycephala*), in different seasons. In Australia foliage gleaning birds that live in the same habitat tend not to differ in the tree species they use, perhaps because eucalypts are rather uniform in their foliage. Bark-feeding birds though differ in their exploitation of different types of bark.

Insectivorous birds appear to avoid or at least reduce competition by diverging in their niches. There is circumstantial evidence that their food is sometimes scarce and that their foraging activities contribute to this scarcity. Nectar-feeders compete more openly, by interspecific aggression and on occasion by depletion of their food. The species that are more dependent on nectar appear to coexist within a region by the larger more aggressive species defending rich nectar sources (unless food is very abundant) and the smaller species exploiting poorer sources, or sneaking into rich areas or swamping them by flocking. This is remarkably similar to the pattern found in hummingbirds in Central America.

Worldwide, bird assemblages are more diverse in more productive habitats and in those with greater structural diversity, though the diversity of plant species may also influence diversity. This is reflected in the rather predictable successional changes that take place after major disturbances like fire and logging.

The theoretical framework of community ecology, constructed largely from studies in the Northern Hemisphere, emerges as very sound after scrutiny from an antipodean perspective. The complex concept of stability, by which I mean persistence and resilience of the community, has not yet received much attention. There is clearly a great need for long-term study of the structure of bird assemblages in Australia.

Breeding Biology

Food is perhaps the universal ultimate factor that determines when birds breed. In most areas, probably including Australia, changes in day-length cause the physiological and behavioural changes that birds must undergo before they can start breeding. The weather can modify the time that breeding is initiated. Cold spells can hold up breeding in the temperate parts of Australia, just as in the Northern Hemisphere. Rain, though, is probably a major modifying factor in northern Australia and in the centre. Opportunistic breeding, especially after rain in inland Australia, has probably been over-emphasized though. Most species in most areas show a marked seasonal peak in breeding in spring to early summer. Autumn breeding is shown frequently in some species, and in some years by many species. It is probably more common in Australia than in Europe and North America. The reason for this is perhaps not because conditions are especially suitable in Australia for breeding in autumn, but more because young birds reared in autumn have a reasonable chance of surviving the relatively mild winter.

Several other features may influence the timing of breeding. Heavy rain in the middle of the wet season in the monsoonal north of Australia and New Guinea may lead to the peak in breeding being just before or just after the wettest months. Nest-robbing reptiles and parasitic cuckoos are more prevalent late in spring, at least in temperate regions. This may encourage birds to breed as early as possible. Many more studies of the breeding biology of Australian birds are required, particularly in the monsoonal north and in the arid centre.

Australian birds have long breeding seasons compared with species at the same latitude in the Northern Hemisphere. Wyndham (1986) showed that this was not because Australia is mostly arid. Woinarski (1985b) noted several other features of Australian birds which differed quantitatively from their northern counterparts. Clutch sizes are smaller in most groups and few species are migratory. Many populations contain non-breeding birds, which frequently help breeding pairs. All of these characteristics can be interpreted as responses to an environment that is not markedly seasonal. Insects in eucalypt woodlands and forests tend not to show great peaks in spring, nor periods of great scarcity in winter. It is perhaps easier to survive the winter in such habitats than in northern deciduous and coniferous forests. On the other hand the lack of a period of great food surplus means that breeding may be difficult and few young can be reared in any brood. Seed-eating birds have large clutches, show more marked seasonal breeding and do not breed co-operatively. Seeds could well show marked seasonal peaks and periods of scarcity. One can link the lack of seasonality in insect availability to the evergreen nature of the vegetation throughout Australia. This presumably results from the rather equable temperature throughout most of the year. Only in a few areas does cold weather halt plant growth entirely in winter. Hot, dry weather may prevent growth but insects still remain active unless drought is extreme. Energy requirements of birds may also be low under these conditions. The equable climate can be traced further back to the fact that land masses in the Southern Hemisphere are smaller than in the north, and are surrounded by large expanses of sea. The latter moderates seasonal extremes.

There is now enough information to propose this link between breeding behaviour of birds and seasonality of vegetation and insects to climate. However, it still remains a tentative hypothesis, which needs to be tested by comparisons between habitats and years and by experiments involving food removal and supplementation. The view of Australia as an equable place is at variance with an earlier prevalent view that it is a harsh continent where the animals need to respond opportunistically. The two views are not incompatible if equable means lack of strong seasonality, and harshness relates to changes between years. Some years, when rainfall is deficient, may be very hard so that breeding is difficult at almost any time and adult survival may be reduced. Drought, in terms of below average rainfall, can affect all habitats and could be as damaging in rainforest as in desert. Few studies have spanned wet and dry periods and so understanding of year-to-year changes in behaviour of birds in response to the weather is based on glimpses of small sections of the overall range. In central Australia it is perhaps the ornithologists as much as the birds that are opportunistic.

By no means do all Australian landbirds breed as simple territorial pairs, which is the rule in most Northern Hemisphere species. Somewhat under 10% of species are polygamous, though the proportion is higher in New Guinea. Most of these species fall into ecological groupings that have been recognized elsewhere. Several occupy simple habitats like grassland and marsh; indeed they are members of the Eurasian and African warbler families (Cisticoliidae, Sylviidae), which have many polygamous species elsewhere. The bowerbirds and birds of paradise are frugivores, mostly dwelling in rainforest and are counterparts of the Neotropical cotingas and manakins. There appear to be no equivalents to the lekking waders and grouse of the north where the male has become emancipated from parental duties because the young are precocial. Perhaps detailed studies of the endemic waders would reveal polygamous behaviour. The lyrebirds appear to be unique in several ways. They feed on arthropods in forest and the single young requires extensive parental care. Despite this they are polygamous with males presenting a most elaborate audiovisual display. Lill (1986) argued that the uncertainty of paternity and the unlikelihood that male parental care could increase reproductive success above the single young per year are the key factors in the evolution of polygamy in lyrebirds. This is speculative, as are the explanations for the use of mimicry in the display song.

Studies on bowerbirds and birds of paradise have extended the theories relating to the evolution of polygamy. By experimentation Borgia (1986) was able to show that female Satin Bowerbirds (*Ptilinorhynchus violaceus*) select males that are genetically superior, as evidenced by their ability to survive a long time and to protect and enhance their bowers. Beehler (1987) showed that the link between polygamy and frugivory is by no means simple. Birds of paradise that eat poor quality fruits tend to be monogamous. A species needs to feed on a wide variety of rich fruits and occupy a large home range for males to have an opportunity to meet and mate with several females.

Perhaps one third of Australian landbirds are co-operative breeders. Co-operative breeding is a fascinating subject because the helpers appear to behave altruistically, i.e., to increase the reproductive success of others at the expense of their own. This has caused difficulties for behavioural ecologists, who attempt to explain behaviour in terms of enhancing individual fitness. Studies on Australian co-operative breeders, particularly those of Rowley, have been at the forefront in explaining the relative benefits of helping behaviour. Helping behaviour is often not an alternative to breeding successfully. Typically helpers have been unable to find a mate or suitable habitat in which to breed. They could either survive in suboptimal habitat and perhaps attempt to breed with little chance of success or remain in a group and help. The second option may be better because group-living offers some protection from predators and usually a more reliable food source. Also, they are in a good position to occupy a vacancy in a good territory when this arises. Further, they are usually helping relatives, typically young males helping their fathers. If they can increase the breeding birds' success they can enhance their own inclusive fitness through kin selection. All of these relative benefits of helping behaviour have also been found in studies in North America, Africa and southern Asia. The ways in which natural selection can favour the evolution of co-operative breeding have now been placed on a sound theoretical base. There is still some dispute, though, over the relative importance of different factors in different species, in particular the necessity for kin selection.

In addition, one of the weaknesses of the explanations for co-operative breeding is that many of the characteristics that supposedly lead to it, e.g., shortage of habitat, presence of surplus non-territorial birds, are also shown by many birds that do not breed co-operatively.

Far more speculative is the ecological basis for co-operative breeding. Why are some species co-operative whereas others, often closely related, are not? Co-operative breeders in Australia, as elsewhere, are a very diverse collection ecologically. This means that a universal explanation for why a species breeds co-operatively is unlikely to be found. The large number of co-operative species in Australia does at least allow some ecological comparisons to be made. Cooperative breeders tend to be insectivorous, to forage on the ground or on otherwise exposed sites, usually by gleaning, and to inhabit open forest or woodland in semi-arid areas. Environments that are not strongly seasonal may lead to more adult birds than there are breeding territories. This is because adult survival is high and as breeding is difficult only some parts of the habitat are suitable. Such a situation is likely to arise in environments with great variation in conditions from year to year, especially when a poor year follows a good one. This habitat saturation theory is already well established. For co-operative breeding to evolve there must be some advantage in staying with the family (or joining a new group) over solitary wandering in search of a territory. I have argued that the advantages of group-living, especially the detection and avoidance of predators, tip the scale in favour of belonging to a group even if this requires helping to rear young that are not your own. Eucalypt woodland is a habitat in which arthropods do not fluctuate greatly between seasons, though there may be variation between years, and in which solitary exposed foragers may be vulnerable to predation. This hypothesis needs to be tested by studies on additional bird species, and also on the patterns of food abundance and the vulnerability of different species to predation.

The mound-builders or megapodes display a unique form of incubation for birds. They use solar, fermentative or volcanic heat to maintain a remarkably constant egg temperature. The young hatch in a very advanced state and require no parental care. Megapodes are endemic to the Australian region and nearby islands and their behaviour extends the variety of breeding biology in birds. It is possible that their ectothermic incubation is primitive, as it resembles that shown by some reptiles. More likely it has evolved secondarily from birds that built more regular nests. Though, ecologically, megapodes are perhaps a curiosity rather than providing a contribution to theory, their embryology may repay detailed study.

Population Ecology

There have been very few long term population studies of Australian birds. This is unfortunate because Australian birds and the environments that they occupy tend to be nearer to their natural state than those in Europe, where most of the best studies have been carried out. Many of the species that have provided the backbone of population ecological theory in birds inhabit artificial, heavily managed or fragmented environments. Predators are often controlled or have been eliminated, artificial nest sites may have been provided and in some cases people provide additional food in winter. Even seabirds are experiencing a radically changed source of food as human fishing has reduced many larger predatory fish species.

Seabirds worldwide have low birth and death rates with delayed maturity. Populations therefore tend not to change greatly over a period of a few years and could be regarded as stable. A frequent pattern though is of a gradual but persistent increase over several decades. This is true, for instance, in both the North Atlantic and Australasian Gannet (*Morus bassana, M. serrator*). As well, seabird populations experience periodic population crashes through disease, exceptional weather or starvation. Gulls in both northern and southern temperate regions have experienced rapid population expansion over recent decades, though this is principally because they exploit mostly land-based food of human origin. Perhaps the cycle of steady increase with periodic abrupt declines is a typical and natural pattern in seabirds. On the other hand such a pattern may represent the upsetting of natural equilibrium by human activities, mostly indirectly. Probably the most important point is that whatever the cause for the dramatic decline, and some in the past have clearly been through human persecution, populations of seabirds have almost without exception steadily recovered. Seabird populations, including those studied in Australia and New Zealand, therefore provide powerful support for the idea that populations are regulated.

There is some support too for regulation, by density dependent means, in Australian landbirds. This seems not to be achieved by high mortality in the non-breeding season. More emphasis is placed on adjustment of birth-rates. Most populations that have been studied have a strongly territorial breeding segment and many surplus individuals. These may be attached to breeding groups as helpers in co-operative species, or belong to a non-breeding flock, as in magpies (*Gymnorhina*) and rosellas (*Platycercus*). Only a proportion of the population breeds so that the number of breeding pairs does not fluctuate as greatly as the non-breeding component. The number of breeding attempts, clutch size and breeding success may vary greatly between years, depending on the prevailing conditions. Thus, Australian studies tend to support the idea that landbirds are regulated by density dependent means. Self-regulation is possibly more prevalent than in the Northern Hemisphere and perhaps the carrying capacity of the environment is more prone to fluctuation between years. This means that although they are regulated, populations are by no means constant.

A few Australian birds, especially in more arid regions, display erratic fluctuations in density. Territorial behaviour, and perhaps density dependent factors, generally are less important. Populations tend to respond to a widely varying environment, and have little influence on the abundance of resources. Such species are far harder to study, and with the exception of the Budgerigar (*Melopsittacus undulatus*), have received little attention. Waterbirds are especially at the mercy of external conditions and many in Australia respond in an opportunistic fashion to rainfall and drought.

Migration

The Australian region has far fewer intercontinental migrants than do other continents. Most are waders or seabirds, and only a handful of landbirds regularly travel between Australia and Asia. This is probably historical, due to Australia's previous isolation, but also ecological as winters are generally not extreme and a wide range of climates is offered within the continent. Most species are sedentary, or at least only wander locally outside the breeding season, even in the southeast and Tasmania. There is a component of regular north–south migrants, mostly birds feeding on large or active insects. Other species migrate altitudinally or inland or towards the coast. There is also a significant group of species that show complex, erratic or nomadic movements. This could be due to the substantial year-to-year variation in conditions, especially in inland Australia. There is a clear need for methodical and long-term studies of these species, from which clearer patterns of movements may emerge.

Priorities for Future Research

Throughout this book, and again in this final chapter I have mentioned areas of study that I see as priorities for future research on Australian bird ecology. Rather than discuss these again I have provided a list (Table 50) of the topics that I believe should receive attention. I have tried to resist the temptation to include everything, because infinitely more remains to be studied than has

Table 50. Research priorities in Bird Ecology in Australia.

1. Long term studies in population densities and the composition of bird assemblages.
2. The influence of changes in human land-use on bird assemblages.
3. Habitat selection with regard to food, other resources and other bird species, especially in rare species.
4. Relationship between behaviour and ecology, especially breeding system in relation to habitat and food.
5. Migratory patterns of both regular migrants and nomads.
6. Foraging niches of birds in a wide range of habitats: desert, grassland, acacia woodland, native pine forest, rainforest, mangrove etc.
7. Relationship between morphology and foraging behaviour, within and between species.
8. The role of birds in regulating insect populations in woodlands and forests.
9. Physiological changes associated with breeding, migration and nomadism, particularly in unpredictable habitats.

already been discovered. In the last few decades the basic distributions of Australian birds have been plotted, outlines of the breeding biology of a diverse variety of species have been described and a reasonable knowledge about bird assemblages of eucalypt forests and woodlands has accumulated. In the future there is a need for expanding in breadth and depth and extending knowledge to a wider range of species and habitats.

I should like to echo Saunders' (1985) comment that far more attention needs to be paid to birds in agricultural and grazing land. A recent book by O'Connor and Shrub (1986) explains how changes in land-use in Britain have had profound and often subtle effects on birds over the last 30 years. These authors had the advantage of long-term studies carried out by the British Trust for Ornithology. There is no comparable data-base for Australia. On the one hand agriculture is becoming more intensive in Australia, whereas, on the other, many farmers are starting to appreciate wildlife. It is also clear that protection, regeneration or even replanting of native vegetation is essential in the battle against erosion and salting of productive soils. It is essential that ecologists work with agricultural scientists so that rural land remains rich in wildlife, at the same time as producing at its maximum potential.

In this book I have attempted to summarize and generalize. Some of the ideas are already well established, others have been developed recently. None should be allowed to become accepted dogma without critical examination. We need to ask specific questions and to test hypotheses rigorously in an attempt to develop theory. There is no reason why the next few decades of research in Australian bird ecology should not be at least as exciting as the previous ones.

Glossary

Altricial: Helpless when hatched.
Arena: Display ground, usually solitary.
Assemblage: Members of a community belonging to one taxon (e.g., birds).
Carrying Capacity: Number of individuals that a habitat can support.
Community: Group of organisms occupying a common environment.
Communal Breeding: Co-operative breeding.
Competition: Use of a common resource that is in short supply, or restriction of access to a resource, even if not in short supply.
Co-operative Breeding: Assistance in breeding by individuals other than parents.
Density-Dependent: Factor whose effect on a population varies with the number of individuals in a population (e.g., food).
Density-Independent: Factor whose effect on the population is not influenced by the number of individuals in a population (e.g., weather).
Diversity: A measure of the number of species in a community and their relative abundance.
Ecotone: The edge between two different habitats.
Endangered: Likely to go extinct if current conditions prevail.
Equable: Environments without harsh conditions in most years.
Exploitation: Competition by depletion of common resources.
Frugivory: Eating fruits.
Granivory: Eating seeds.
Group Selection: Natural selection operating at the level of the species or population.
Guild: A group of ecologically similar species in a community.
Helper: Individual that assists at the nest of others.
Herbivory: Eating live plant material.
Inclusive Fitness: Reproductive success of an individual plus that of relatives corrected for the degree of relatedness.
Individual Selection: Natural selection operating on the individual to enhance its reproductive success.
Insectivory: Eating insects (and usually other arthropods).
Interference: Preventing access to resources, usually by other species.
'K' selection: Tendency to divert resources into survival, with modest reproductive effort.
Kin Selection: Natural selection that results in an increase in the reproductive success of a relative, while at the same time slightly reducing the individual's reproductive success.
Lek: Communal display ground where males display to attract females.
Migration: Regular seasonal change in distribution of a population.
Monogamy: Bonding between a single male and a single female.
Mortality: Death rate.

Mutualism: Interaction between two species from which both benefit.
Natality: Birth rate.
Nectarivory: Eating nectar.
Nidicolous: Remaining in the nest after hatching.
Nidifugous: Leaving the nest soon after hatching.
Nomadism: Irregular change in the distribution of a population.
Parasitism: Living on or in another organism, usually without killing it.
Persistence: Lack of change in a community over a period of time.
Polyandry: Breeding system in which a single female regularly mates with several males.
Polygamy: Breeding system in which one individual of one sex regularly mates with several individuals of the other sex.
Polygyny: Breeding system in which a single male regularly mates with several females.
Precocial: Active soon after hatching.
Predation: Killing and consuming prey.
Predictability: Constancy of conditions from day to day or year to year.
Promiscuity: Mating of individuals with more than one individual of the opposite sex, with no pair-bonding.
Proximate factors: Factors to which an animal responds physiologically in the short-term.
'r' selection: Tendency to divert resources into reproduction at the expense of survival.
Raptor: Bird that frequently preys on other vertebrates.
Recruitment: Addition to the population, often of a particular group, e.g., breeding population.
Resilience: Ability to recover from a disturbance.
Resistance: Ability to withstand a disturbance without changing.
Seasonal: Environments with marked and regular changes in climate through the year.
Stability: A concept with various meanings in ecology, but usually implying low levels of change or resistance to change.
Structure: The composition of a community and the interactions between the species.
Succession: Progressive change in a community in composition and structure.
Territory: A defended area or other resource (e.g., food).
Ultimate factors: Factors that determine the value of a behaviour in the evolutionary sense.
Vulnerable: Likely to become endangered if current conditions prevail.

Bibliography

Abbott, I. 1973. Birds of Bass Strait. Evolution and ecology of the avifaunas of some Bass Strait Islands, and comparisons with those of Tasmania and Victoria. *Proc. Roy. Soc. Vic.* 85: 197–224.

———. 1974. The avifauna of Kangaroo Island and causes of its impoverishment. *Emu* 74: 124–34.

———. 1978. Factors determining the number of land bird species on islands around south-western Australia. *Oecologia* 33: 221– 33.

Alatalo, R. V., Lundberg, A. and Ulfstrand, S. 1985. Habitat selection in the Pied Flycatcher *Ficedula hypoleuca.* In *Habitat Selection in Birds,* ed. M. L. Cody, pp. 59–85. New York: Academic Press.

Amadon, D. 1950. The Hawaiian Honeycreepers (Aves: Drepaniidae). *Bull. Amer. Mus. Nat. Hist.* 95: 151–262.

Ambrose, S. J. 1984. The response of small birds to extreme heat. *Emu* 84: 242–43.

Ambrose, S. J. and Bradshaw, S. D. 1988a. Seasonal changes in standard metabolic rates in the White-browed Scrub-wren *Sericornis frontalis* (Acanthizidae) from arid, semi-arid and mesic environments. *Comp. Biochem. Physiol.* 89A: 79–83.

———. 1988b. The water and electrolyte metabolism of free-ranging and captive White-browed Scrub-wrens *Sericornis frontalis* (Acanthizidae) from arid, semi-arid and mesic environments. *Aust. J. Zool.* 36: 29–51.

Andrén, H., Angelstam, P., Lindström, E. and Widén, P. 1985. Differences in predation pressure in relation to habitat fragmentation: an experiment. *Oikos* 45: 273–77.

Andrewartha, H. G. and Birch, L. C. 1954. *The distribution and abundance of animals.* Chicago: Chicago Univ. Press.

———. 1985. *The Ecological Web: more on the distribution and abundance of animals.* Chicago: Chicago Univ. Press.

Anon, 1988. Chatham Island Black Robin. *RAOU Newsletter* 76: 10.

Ashmole, N. P. 1963. The regulation of numbers of tropical oceanic birds. *Ibis* 103: 458–73.

Attwood, R. 1982. Birds sheltering in the burrows of mammals during a heatwave. *S. Aust. Orn.* 29: 7–9.

Audley-Charles, M. G. 1983. Reconstruction of eastern Gondwanaland. *Nature* 306: 48–50.

Baird, R. F. 1984. The Pleistocene distribution of the Tasmanian Native-hen *Gallinula mortierii mortierii. Emu* 84: 119–23.

Baker, A. J. 1980. Morphometric differentiation in New Zealand populations of the House Sparrow (*Passer domesticus*). *Evolution* 34: 638–53.

Baker, J. R., Marshall, A. J. and Harrisson, T. H. 1940. The seasons in a tropical rainforest (New Hebrides). Part 5. Birds (*Pachycephala*). *J. Linn. Soc. Lond.* 41: 50–71.

Barlow, B. A. 1981. The Australian flora: its origin and evolution. In *Flora of Australia, Volume 1 Introduction,* pp. 25–76. Canberra: Bureau of Flora and Fauna.

Beauchamp, A. J. and Worthy, T. H. 1988. Decline in distribution of the takahe (*Porphyrio* = *Notornis mantelli*): a re-examination. *J. Roy. Soc. N.Z.* 18: 103–12.

Beehler, B. 1983a. For gaudy display, it's hard to beat birds of paradise. *Smithsonian* 13: 91–97.

———. 1983b. Frugivory and polygamy in birds of paradise. *Auk.* 100: 1–12.

———. 1983c. Notes on the behaviour and ecology of Macgregor's Bird of Paradise. *Emu* 83: 28–30.

———. 1987. Birds of paradise and mating system theory — predictions and observations. *Emu* 87: 78–89.

Beehler, B. M., Pratt, T. K. and Zimmerman, D. A. 1986. *Birds of New Guinea.* Princeton, New Jersey: Princeton Univ. Press.

Beehler, B. and Pruett-Jones, S. G. 1983. Display, dispersion and diet of birds of paradise: a comparison on nine species. *Behav. Ecol. Sociobiol.* 13: 229–38.

Beeton, R. J. S. 1975. Corella and grain sorghum: finding a solution. In *Agriculture, Forestry and Wildlife: Conflict or Coexistence?,* ed. P. J. Jarman, pp. 147–62. Armidale: UNE Press.

———. 1977. The impact and management of birds on the Ord River development in Western Australia. M. Nat. Res. thesis, Univ. of New England, Armidale.

Begon, M. 1979. *Investigating animal abundance: Capture-recapture for biologists.* London: Edward Arnold.

Bell, H. L. 1965. A population study of heathland birds. *Emu* 65: 295–304.

———. 1979. The effects on rainforest birds of planting of Teak *Tectona grandis* in Papua New Guinea. *Aust. Wildl. Res.* 6: 305–18.

———. 1980. Composition and seasonality of mixed-species feeding flocks of insectivorous birds in the Australian Capital Territory. *Emu* 80: 227–32.

———. 1982a. Sexual differences in the foraging behaviour of the frill-necked flycatcher *Arses telescopthalmus* in New Guinea. *Aust. J. Ecol.* 7: 137–47.

———. 1982b. A bird community of lowland rainforest in New Guinea. 1. Composition and density of the avifauna. *Emu* 82: 24–41.

———. 1982c. A bird community of lowland rainforest in New Guinea. 2. Seasonality. *Emu* 82: 65–74.

———. 1982d. Abundance and seasonality of the savanna avifauna at Port Moresby, Papua New Guinea. *Ibis* 124: 252–71.

———. 1983a. Resource-partitioning among three syntopic thornbills (Aves: *Acanthiza* Vigors and Horsfield). Ph.D. thesis, Univ. of New England, Armidale.

———. 1983b. A bird community of lowland rainforest in New Guinea. 5. Mixed species flocks of insectivorous birds. *Emu* 82: 256–75.

———. 1984. A bird community of lowland rainforest in New Guinea. 6. Foraging ecology and community structure of the avifauna. *Emu* 84: 142–58.

———. 1985a. Seasonal variation and the effects of drought on the abundance of arthropods in savanna woodland on the Northern Tablelands of New South Wales. *Aust. J. Ecol.* 10: 207–22.

———. 1985b. The social organization and foraging behaviour of three synoptic thornbills *Acanthiza* spp. In *Birds of eucalypt forests and woodlands: ecology, conservation, management,* eds. A. Keast, H. F. Recher, H. A. Ford and D. Saunders, pp. 151–163. Sydney: RAOU, Surrey Beatty.

———. 1986a. Sexual differences in the behaviour of wintering Golden Whistlers *Pachycephala pectoralis* at Wollomombi, NSW. *Emu* 86: 2–11.

———. 1986b. The participation by cuckoos in mixed-species flocks of insectivorous birds in south-eastern Australia. *Emu* 86: 249–53.

———. 1986c. Occupation of urban habitats by birds in Papua New Guinea. *Western Foundation of Vertebrate Zoology* 3: 1–48.

Bell, H. L. and Ferrier, S. 1985. The reliability of estimates of density from transect counts. *Corella* 9: 3–13.

Bell, H. L. and Ford, H. A. 1986. A comparison of the social organization of three syntopic species of Australian thornbill *Acanthiza. Behav. Ecol. Sociobiol.* 19: 381–92.

———. 1987. Fidelity to banding-site in four migratory species of birds at Wollomombi, N.S.W. *Corella* 11: 1–5.

Bell, H. L. and Swainson, G. W. 1985. The colonization, ecology and breeding of the Pied Stonechat *Saxicola caprata* at Port Moresby, Papua New Guinea. *Ibis* 127: 74–83.

Birkhead, T. R., Atkin, L. and Møller, A. P. 1987. Copulation behaviour of birds. *Behaviour* 100: 101–38.

Blakers, M., Davies, S. J. J. F. and Reilly, P. N. 1984. *The atlas of Australian Birds.* Melbourne: Melbourne Univ. Press.

Blondel, J. 1985. Breeding strategies of blue tit and coal tit (*Parus*) in mainland and island Mediterranean habitats. *J. Anim. Ecol.* 54: 531–56.

Booth, D. T. and Seymour, R. S. 1984. Effect of adding water to Mallee Fowl mounds during drought. *Emu* 84: 116–18.

Borgia, G. 1985a. Bower destruction and sexual competition in the Satin Bowerbird *Ptilinorhynchus violaceus. Behav. Ecol. Sociobiol.* 18: 91–100.

———. 1985b. Bower quality, number of decorations and mating success of male Satin Bowerbirds (*Ptilinorhynchus vidaceus*): An experimental analysis. *Anim. Behav.* 33: 266–71.

———. 1986. Sexual selection in Bowerbirds. *Scientific American* 254: 92–100.

Borgia, G., Pruett-Jones, S. G. and Pruett-Jones, M. A. 1985. The evolution of bower-building and the assessment of male quality. *Z. Tierpsychol.* 67: 225–36.

Boswall, J. 1978. Further notes on tool-using by birds and related behaviour. *Avicult. Mag.* 84: 162–66.

Bowman, R. I. and Billeb, S. L. 1965. Blood-eating in a Galapagos finch. *Living Bird* 4: 29–44.

Braithwaite, L. W. 1976. Breeding seasons of waterfowl in Australia. *Proc. 16th Int. Orn. Congress.* (*Canberra.*) pp. 235–47.

Breckwoldt, R. 1983. *Wildlife in the home paddock. Nature conservation for Australian farmers.* Sydney: Angus and Robertson.

Brereton, J. LeG. 1963. The life cycles of three Australian parrots: some comparative and population aspects. *Living Bird* 2: 21–9.

———. 1971. Inter-animal control of space: the role of rank order, parent-offspring relations and peer play, illustrated by comparative studies of Australian parrots. In *Behaviour and environment*, ed. A. Esser, New York: Plenum.

Bridges, L. 1980. Some examples of the behaviour and feeding ecology of the Rufous (*Pachycephala rufiventris*) and Golden (*P. pectoralis*) Whistler. Unpubl. B.Sc. (Hons) thesis, Univ. of New England, Armidale.

Bridgewater, P. 1987. The present Australian environment. In *Fauna of Australia. Volume 1A. General Articles,* ed. G. R. Dyne, pp. 69–100. Canberra: Australian Government.

Briggs, S. V. 1977. Variation in waterbird numbers at four swamps on the Northern Tablelands of New South Wales. *Aust. Wildl. Res.* 4: 301–9.

Brooker, M. G., Ridpath, M. G., Estbergs, A. J., Bywater, J., Hart, D. S. and Jones, M. S. 1979. Bird observations on the north-western Nullarbor Plain and neighbouring regions, 1967–1978. *Emu* 79: 176–90.

Brown, J. L. 1964. The evolution of diversity in avian territorial systems. *Wilson Bull.* 76: 160–69.

———. 1978. Avian communal breeding systems. *Ann. Rev. Ecol. Syst.* 9: 123–56.

———. 1980. Fitness in complex avian social systems. In *Evolution of social behaviour: hypothesis and empirical tests,* ed. H. Markl, Weinheim: Verlag Chemie.

———. 1987. *Helping and communal breeding in birds: ecology and evolution.* Princeton, N. J.: Princeton Univ. Press.

Brown, J. L., Brown, E. R., Brown, S. D. and Dow, D. D. 1982. Helpers — effects of experimental removal on reproductive success. *Science* 215: 421–22.

Brown, J. L., Dow, D. D., Brown, E. R. and Brown, S. D. 1978. Effects of helpers on feeding of nestlings of Grey-crowned Babbler *(Pomatostomus temporalis). Beh. Ecol. Sociobiol.* **4:** 43–59.

Brown, P. B. and Wilson, R. I. 1982. The Orange-bellied Parrot. In *Species at risk: research in Australia,* eds. R. H. Groves and W. D. L. Ride, pp. 106–115. Canberra: Aust. Acad. Science.

Brown, P., Wilson, R., Loyn, R. and Murray, N. 1985. The Orange-bellied Parrot — an RAOU Conservation Statement. *RAOU Report No 14,* Melbourne.

Bruns, H. 1960. The economic importance of birds in forests. *Bird Study* 7: 193–208.

Buckland, S. T., Rowley, I. and Williams, D. A. 1983. Estimation of survival from repeated sightings of tagged Galahs. *J. Anim. Ecol.* 52: 563–73.

Buller, W. L. 1888. *A history of birds of New Zealand.* London.

Burbidge, A. A. and Fuller, P. J. 1982. Banded Stilt breeding at Lake Barlee, Western Australia. *Emu* 82: 212–16.

Burton, T. C. and Martin, A. A. 1976. Analysis of hybridization between Black-backed and White-backed Magpies in south-eastern Australia. *Emu* 76: 30–6.

Cain, A. J. 1955. A revision of *Trichoglossus haematodus* and of the Australian platycercine parrots. *Ibis* 97: 432–79.

Cameron, E. 1985. Habitat usage and foraging behaviour of three fantails (*Rhipidura:* Pachycephalidae). In *Birds of eucalypt forests and woodlands: Ecology, behaviour and conservation.* pp. 177–91. *See* Bell 1985b.

Cannon, C. E. 1981. The diet of Eastern and Pale-headed Rosellas. *Emu* 81: 101–10.

Carpenter, F. L. 1978. A spectrum of nectar-eater communities. *Amer. Zool.* 18: 809–19.

Carpenter, F. L. and MacMillen, R. E. 1976. Threshold model of feeding territoriality and test with a Hawaiian honeycreeper. *Science* 194: 639–42.

Carrick, R. 1972. Population ecology of the Australian Black-backed Magpie, Royal Penguin and Silver Gull. In *Population ecology of migratory birds: a symposium.* U.S. Dept. Interior. pp. 41–99.

Catchpole, C. K. 1980. Sexual selection and the evolution of complex songs among European warblers of the genus *Acrocephalus. Behaviour* 74: 149–66.

———. 1986. Song repertoires and reproductive success in the Great Reed Warbler *Acrocephalus arundinaceus. Beh. Ecol. Sociobiol.* 19: 439–45.

Catterall, C. P. 1985. Winter energy deficits and the importance of fruit versus insects in a tropical island bird population. *Aust. J. Ecol.* 10: 265–79.

Catterall, C. P., Wyatt, W. S. and Henderson, L. J. 1982. Food resources, territory density and reproductive success of an island Silvereye population *Zosterops lateralis*. *Ibis* 124: 405–21.

Chisholm, A. H. 1953. Polygamy among ground-frequenting birds. *Emu* 53: 88–89.

———. 1960. *The Romance of the Lyrebird,* Sydney: Angus and Robertson.

Christensen, P. E. S., Wardell-Johnson, G. and Kimber, P. 1985. Birds and fire in south-western forests. In *Birds of eucalypt forests and woodlands: Ecology, conservation, management,* pp. 291–99. *See* Bell 1985b.

Churchill, D. M. and Christensen, P. 1970. Observations on pollen-harvesting by brush-tongued lorikeets. *Aust. J. Zool.* 18: 427–37.

Clark, L. R. 1964. Predation of birds in relation to the population density of *Cardiaspina albitextura* (Psyllidae). *Aust. J. Zool.* 12: 329–61.

Clarke, M. F. 1984. Cooperative breeding by the Australian Bell Miner *Manorina melanophrys* Latham: A test of kin selection theory. *Behav. Ecol. Sociobiol.* 14: 137–46.

———. 1988. The reproductive behaviour of the Bell Miner *Manorina melanophrys*. *Emu* 88: 88–100.

Cleland, J. B. 1911–12. Examination of contents of stomachs and crops of Australian birds. *Emu* 11: 79–95. *Emu* 12: 8–18.

———. 1919. The birds of the Pilliga Scrub, New South Wales. *Emu* 19: 272–85.

Cody, M. L. 1966. A general theory of clutch size. *Evolution* 20: 174–84.

———. 1974. *Competition and the structure of bird communities.* Princeton, New Jersey: Princeton Univ. Press.

Collins, B. G. and Briffa, P. 1982. Seasonal variation of abundance and foraging of three species of Australian honeyeaters. *Aust. Wildl. Res.* 9: 557–69.

———. 1983. Seasonal and diurnal variations in the energetics and foraging activities of the brown honeyeater *Lichmera indistincta*. *Aust. J. Ecol.* 8: 103–11.

Collins, B. G., Wykes, B. J. and Nichols, O. G. 1985. Re-colonization of restored bauxite minelands by birds in south-western Australian forests. In *Birds of eucalypt forests and woodlands: Ecology, conservation, management,* pp. 341–54. *See* Bell 1985b.

Connell, J. H. 1983. On the prevalence and relative importance of interspecific competition: evidence from field experiments. *Am. Nat.* 122: 661–96.

Considine, M. 1986. Black-eared Miners — a forgotten species. RAOU Newsletter 68: 1–2.

Coulson, J. C., Potts, G. R., Deans, I. R. and Fraser, S. M. 1968. Exceptional mortality of shags and other seabirds caused by paralytic shellfish poisoning. *British Birds* 9: 381–404.

Counsilman, J. J. 1980. A comparison of two populations of the Grey-crowned Babbler. (Part 2). *Bird Behaviour* (Part 2) 2: 1–111.

Craig, J. L. and Douglas, M. E. 1984. Temporal partitioning of a nectar resource in relation to competitive asymmetries. *Anim. Behav.* 32: 624–25.

Crawford, A. M. and Kalmakoff, J. 1977. A host-virus interaction in a pasture habitat *Wiseana* spp. (Lepidoptera: Hepialidae) and its Baculoviruses. *J. Invert. Pathology* 29: 81–7.

Crome, F. H. J. 1978. Foraging ecology of an assemblage of birds in lowland rainforest in northern Queensland. *Aust. J. Ecol.* 3: 195–212.

Crome, F. H. J. and Brown, H. E. 1979. Notes on social organization and breeding of the Orange-footed Scrubfowl *Megapodius reinwardt*. *Emu* 79: 111–19.

Dann, P. 1987. The feeding behaviour and ecology of shorebirds. In *Shorebirds in Australia,* pp. 10–21. *See* Lane 1987.

———. 1988. An experimental manipulation of clutch size in the Little Penguin *Eudyptula minor*. *Emu* 88: 101–04.

Darwin, C. 1871. *The Descent of Man, and selection in relation to sex*. London: J. Murray.

Davies, N. B. 1978. Ecological questions about territorial behaviour. In *Behavioural Ecology: an evolutionary approach,* eds. J. R. Krebs and N. B. Davies, pp. 317–350. London: Blackwell.

Davies, S. J. J. F. 1977. The timing of breeding by the Zebra Finch *Taeniopygia castanotis* at Mileura, Western Australia. *Ibis* 119: 369–72.

———. 1984. Nomadism as a response to desert conditions in Australia. *J. Arid Environments* 7: 183–95.

Davis, W. E. 1986. A White-faced Heron *Ardea novaehollandiae* uses possible anti-glare foraging behaviour. *Corella* 10: 65–6.

Dawson, W. R. and Bartholomew, G. A. 1968. Temperature regulation and water economy of desert birds. In *Desert Biology,* ed. G. W. Brown, pp. 357–95. New York: Academic Press.

Debus, S. J. S. and Czechura, G. V. 1988. The Red Goshawk *Erythrotriorchis radiatus:* a review. *Australian Bird-Watcher* 12: 175–99.

Delroy, L. 1985. Penguin deaths in the south-east of South Australia. *S. Aust. Orn.* 29: 194.

Diamond, J. M. 1975a. Assembly of Species Communities. In *Ecology and Evolution of Communities,* eds. M. L. Cody and J. M. Diamond, pp. 342–444. Cambridge, Mass: Belknap.

———. 1975b. The island dilemma: lessons of modern biogeographic studies for the design of nature reserves. *Biol. Conserv.* 7: 129–46.

———. 1982. Evolution of bowerbirds' bowers: animal origins of the aesthetic sense. *Nature* 297: 99–102.

———. 1986. Evolution of ecological segregation in the New Guinea montane avifauna. In *Community Ecology,* eds. J. M. Diamond and E. J. Case, pp. 98–125. New York: Harper and Row.

Diamond, J. M. and Veitch, C. R. 1981. Extinctions and introductions in the New Zealand avifauna: Cause and Effect? *Science* 211: 499–501.

Dick, W. J. A., Piersma, T. and Prokosch, P. 1987. Spring migration of the Siberian Knots *Calidris canutus canutus*: results of a co-operative Wader Study Group project. *Ornis Scand.* 18: 5–16.

Dow, D. D. 1970. Communal behaviour of nesting Noisy Miners. *Emu* 70: 131–34.

———. 1977. Indiscriminate interspecific aggression leading to almost sole occupancy of space by a single species of bird. *Emu* 77: 115–21.

———. 1978a. Breeding biology and development of the young of *Manorina melanocephala,* a communally breeding honeyeater. *Emu* 78: 207–22.

———. 1978b. Systems and strategies of communal breeding in Australian birds. Proc. 17th Int. Ornithol. Congr. (Berlin), pp. 875–81.

———. 1980. Communally breeding Australian birds with an analysis of distributional and environmental factors. *Emu* 80: 121–40.

Driscoll, P. V. 1977. Comparison of bird counts from pine forests and indigenous vegetation. *Aust. Wildl. Res.* 4: 281–88.

Dwyer, P. D. 1972. Feature, patch and refuge area: some influences on diversity of bird species. *Emu* 72: 149–56.

Ehrlich, P. R., Holm, R. W. and Parnell, D. R. 1974. *The Process of Evolution.* Tokyo: McGraw-Hill.

Elton, C. S. 1958. *The ecology of invasions by animals and plants.* London: Methuen.

Emison, W. B., Du Guesclin, P. B. and Temby, I. D. 1981. Management of the Long-billed Corella. In *Wildlife Management in the '80s,* ed. T. Riney, pp. 248–57. Melbourne: Monash Univ.

Emlen, J. T. 1971. Population densities of birds derived from transect counts. *Auk* 88: 323–42.

Emlen, S. T. 1978. The evolution of co-operative breeding in birds. In *Behavioural ecology: an evolutionary approach,* pp. 245–81. *See* Davies 1978.

———. 1982a. The evolution of helping. I. An ecological constraints model. *Am. Nat.* 119: 29–39.

———. 1982b. The evolution of helping. II. The role of behavioural conflict. *Am. Nat.* 119: 40–53.

Emlen, S. T. and Oring, L. W. 1977. Ecology, sexual selection and the evolution of mating systems. *Science* 197: 215–23.

Emlen, S. T. and Vehrencamp, S. L. 1985. Co-operative breeding strategies among birds. In *Experimental Behavioural Ecology,* eds. Hölldobler and Lindauer. Stuttgart: Verlag.

Endler, J. A. 1977. *Geographic variation, speciation and clines.* Princeton, New Jersey: Princeton Univ. Press.

Enemar, A., Nilsson, L. and Sjöstrad, B. 1984. The composition and dynamics of the passerine bird community in a subalpine birch forest, Swedish Lapland. A 20-year study. *Ann. Zool. Fennici* 21: 321–38.

Entwistle, P. F., Adams, P. H. W. and Evans, H. F. 1977. Epizootiology of a Nuclear-Polyhedrosis virus in European Spruce Sawfly, *Gilpinia hercyniae*: Birds as dispersal agents of the virus during winter. *J. Invert. Pathology* 30: 15–9.

Erdelen, M. 1984. Bird communities and vegetation structure. 1. Correlations and comparisons of simple and diversity indices. *Oecologia* 61: 277–84.

Feinsinger, P. 1976. Organization of a tropical guild of nectarivorous birds. *Ecol. Monogr.* 46: 257–91.

Ferrier, S. 1985. Habitat requirements of a rare species, the Rufous Scrub-bird. In *Birds of eucalypt forests and woodlands: ecology, management, conservation,* pp. 241–248. *See* Bell 1985b.

Fisher, R. A. 1930. *The genetical theory of natural selection.* Oxford: Clarendon.

Fjeldså, J. 1985. Classification of waterbird communities in south-eastern Australia. *Emu* 85: 141–49.

Flack, J. A. D. 1977. Interisland transfers of New Zealand Black Robins. In *Endangered Birds,* ed. S. A. Temple, pp. 365–72. London: Croom Helm.

Fleay, D. H. 1937. Nesting habits of the Brush Turkey. *Emu* 36: 153–63.

Fleming, P. 1980. A comparative ecology of four sympatric robins. Hons. thesis, Univ. of New England, Armidale.

Fogden, M. F. L. 1972. The seasonality and population dynamics of equatorial forest birds in Sarawak. *Ibis* 114: 307–44.

Ford, H. A. 1978. The Black Honeyeater: nomad or migrant? *S. Aust. Orn.* 27: 263–69.

———. 1979. Interspecific competition in Australian honeyeaters — depletion of common resources. *Aust. J. Ecol.* 4: 145–64.

———. 1980. Breeding and moult in honeyeaters (Aves: Meliphagidae) near Adelaide, South Australia. *Aust. Wildl. Res.* 7: 453–63.

———. 1983. Relation between number of honeyeaters and intensity of flowering near Adelaide, South Australia. *Corella* 7: 25–31.

———. 1985a. A synthesis of the foraging ecology and behaviour of birds in eucalypt forests and woodlands. In *Birds of eucalypt forests and woodlands: ecology, conservation, management,* pp. 249–54. *See* Bell 1985b.

———. 1985b. Nectarivory and pollination by birds in southern Australia and Europe. *Oikos* 44: 127–31.

———. 1985c. Nectar-feeding birds and bird-pollination: why are they so prevalent in Australia yet absent from Europe. *Proc. Ecol. Soc. Aust.* 14: 153–58.

———. 1985d. The bird community in eucalypt woodland and eucalypt dieback in the Northern Tablelands of New South Wales. In *Birds of eucalypt forests and woodlands: ecology, conservation, management,* pp. 333–40. *See* Bell 1985b.

———. 1986a. Birds and eucalypt dieback in north-eastern New South Wales. In *The dynamic partnership: birds and plants in southern Australia,* eds. H. A. Ford and D. C. Paton, pp. 150–54. Adelaide: Govt. Printer.

———. 1986b. Concluding remarks: germination of new ideas or flights or fancy? In *The dynamic partnership: birds and plants in southern Australia,* pp. 155–158. *See* Ford 1986a.

Ford, H. A. and Bell, H. L. 1982. Density of birds in eucalypt woodland affected to varying degrees by dieback. *Emu* 82: 202–8.

Ford, H. A., Bell, H., Nias, R. and Noske, R. 1988. The relationship between ecology and the incidence of co-operative breeding in Australian birds. *Beh. Ecol. Sociobiol.* 22: 239–49.

Ford, H. A., Bridges, L. and Noske, S. 1985. Density of birds in eucalypt woodland near Armidale, north-eastern New South Wales. *Corella* 9: 97–107.

Ford, H. A. and Howe, R. W. 1980. The future of birds in the Mount Lofty Ranges. *S. Aust. Orn.* 28: 85–9.

Ford, H. A., Huddy, L. and Bell, H. 1989. Seasonal changes in foraging behaviour of three passerines in Australian eucalypt woodland. *Studies in Avian Biology,* in press.

Ford, H. A., Noske, S. and Bridges, L. 1986. Foraging of birds in eucalypt woodland in north-eastern New South Wales. *Emu* 86: 169–79.

Ford, H. A. and Paton, D. C. 1976a. Resource partitioning and competition in honeyeaters of the genus *Meliphaga. Aust. J. Ecol.* 1: 281–87.

———. 1976b. Impoverishment of the avifauna of Kangaroo Island. *Emu* 76: 155–56.

———. 1977. The comparative ecology of ten species of honeyeaters in South Australia. *Aust. J. Ecol.* 2: 399–407.

———. 1982. Partitioning of nectar sources in an Australian honeyeater community. *Aust. J. Ecol.* 7: 149–59.

———. 1985. Habitat selection in Australian honeyeaters, with special reference to nectar productivity. In *Habitat Selection in Birds,* pp. 367–88. *See* Alatalo *et al.* 1985.

———. 1986. *The dynamic partnership: birds and plants in Southern Australia.* Adelaide: Govt. Printer.

Ford, H. A. and Pursey, J. F. 1982. Status and feeding of the eastern spinebill *Acanthorhynchus tenuirostris* at New England National Park, north-eastern N.S.W. *Emu* 82: 203–11.

Ford, J. R. 1982. Origin, evolution and speciation of birds specialized to mangroves in Australia. *Emu* 82: 12–23.

———. 1983. Evolutionary and ecological relationships between quail-thrushes. *Emu* 83: 152–72.

Ford, J. R. and Johnstone, R. E. 1983. The Rusty-tailed Flyeater, a new species from Queensland. *W. Aust. Nat.* 15: 133–35.

Fordyce, J. C. 1978. Flame Robin banding in the Shepparton district, Victoria. *Corella* 2: 85–8.

Forshaw, J. M. 1981. *Australian Parrots*. Melbourne: Landsdowne.
Franklin, D., Menkhorst, P. and Robinson, J. 1987. Field Surveys of the Regent Honeyeater *Xanthomyza phrygia* in Victoria. *Aust. Bird Watcher* 12: 91–5.
Friend, G. F. 1982. Bird populations in exotic pine plantations and indigenous eucalypt forests in Gippsland, Victoria. *Emu* 82: 80–91.
Frith, D. W. 1984. Foraging ecology of birds in an upland tropical rainforest in North Queensland. *Aust. Wildl. Res.* 11: 325–47.
Frith, H. J. 1956. Breeding of the Mallee Fowl *Leipoa ocellata* Gould. *CSIRO Wildl. Res.* 4: 31–60.
———. 1957. Clutch size of the Goldfinch. *Emu* 57: 287-88.
———. 1962a. *The Mallee-Fowl: the bird that builds an incubator*. Sydney: Angus and Robertson.
———. 1962b. Movements of the Grey Teal *Anas gibberifrons* Müller (Anatidae). *CSIRO Wildl. Res.* 7: 50–70.
———. 1967. *Waterfowl in Australia*. Sydney: Angus and Robertson.
———. 1976. *Reader's Digest complete book of Australian birds*. Sydney: Reader's Digest.
———. 1982. *Pigeons and doves of Australia*. Adelaide: Rigby.
———. 1985. Megapode. In *A dictionary of birds,* eds. B. Campbell and E. Lack, pp. 344–45. Calton, Staffordshire: Poyser.
Frith, H. J. and Waterman, M. H. 1977. Movements of Stubble Quail, *Coturnix pectoralis* from South Australian grainfields. *Aust. Wild. Res.* 4: 85–90.
Fry, C. H. 1977. The evolutionary significance of co-operative breeding in birds. In *Evolutionary Ecology* pp. 127-35, eds. B. Stonehouse and C. M. Perrins. London: MacMillan.
Fullagar, P. J. 1976. Cabbage Tree Island, N.S.W. *Aust. Bird Bander* 14: 94–7.
Fullagar, P. J., Lowe, K. W. and Davies, S. J. J. F. 1988. Intracontinental migration in Australian birds. Proc. 19th Int. Orn. Soc. (Ottawa): in press.
Fuller, R. J. 1982. *Bird habitats in Britain*. Calton, Staffordshire: Poyser.
Fuller, R. J. and Glue, D. E. 1977. The breeding biology of the Stonechat and Whinchat. *Bird Study* 24: 215–28.
Garcia, E. 1983. An experimental test of competition for space between Blackcaps *Sylvia atricapilla* and Garden Warblers *Sylvia borin* in the breeding season. *J. Anim. Ecol.* 52: 795–805.
Garnett, S. 1985. Mortality and group cohesion in migrating Rainbow Bee-eaters. *Emu* 85: 267–68.
Gaston, A. J. 1977. Social behaviour within groups of Jungle Babblers *Turdoides striatus*. *Anim. Behav.* 25: 828–48.
———. 1978. The evolution of group territorial behaviour and co-operative breeding. *Am. Nat.* 112: 1091–100.
———. 1985. Co-operative breeding. In *A dictionary of birds*. pp. 108–9. See Frith 1985.
Gepp, B. C. 1986. Birds in pine forests in South Australia. In *The dynamic partnership: birds and plants in southern Australia,* pp. 140–49. See Ford 1986a.
Gibb, J. A. 1950. The breeding biology of Great and Blue Titmice. *Ibis* 92: 507–39.
Gibson, J. D. 1979. Growth in the population of the Silver Gull on the Five Islands Group, New South Wales. *Corella* 3: 103–4.
Gill, F. B. and Wolf, L. L. 1975. Economics of feeding territoriality in the golden-winged sunbird. *Ecology* 56: 333–45.
Giller, P. S. 1984. *Community Structure and the Niche*. London: Chapman and Hall.
Gilliard, E. T. 1969. *Birds of Paradise and Bower Birds*. New York: Nat. History Press.

Gilmore, A. M. 1985. The influence of vegetation structure on the density of insectivorous birds. In *Birds of eucalypt forests and woodlands: ecology, conservation, management,* pp. 21–31. *See* Bell 1985b.

Grant, P. R. 1986. *Ecology and evolution of Darwin's Finches.* Princeton, New Jersey: Princeton Univ. Press.

Green, R. H. 1966. Gut contents of some Tasmanian birds. *Emu* 66: 105–10.

Green, R. J. 1984. Native and exotic birds in a suburban habitat. *Aust. Wild. Res.* 11: 181–90.

———. 1986. Native and exotic birds in the suburban habitat. In *The dynamic partnership: birds and plants in southern Australia,* pp. 130–39. *See* Ford 1986a.

Greig-Smith, P. W. 1978. The formation, structure and function of mixed-species insectivorous bird flocks in Western African savanna woodland. *Ibis* 120: 284–97.

Grimes, L. G. 1976. The occurrence of co-operative breeding behaviour in African birds. *Ostrich* 47: 1–15.

Grinnell, J. 1904. The origin and distribution of the Chestnut-backed Chickadee. *Auk* 21: 364–82.

Halliday, T. R. 1978. Sexual selection and mate choice. In *Behavioural ecology: an evolutionary approach,* pp. 180–213. *See* Davies 1978.

———. 1980a. *Sexual Strategy.* Oxford: Oxford Univ. Press.

———. 1980b. *Vanishing Birds: their natural history and conservation.* England: Penguin.

Harden, R. H., Muir, R. J. and Milledge, D. R. 1986. An evaluation of the strip transect method for censusing bird communities in forests. *Aust. Wildl. Res.* 13: 203–11.

Hardin, G. 1960. The competitive exclusion principle. *Science* 131: 1292–97.

Harris, M. P. 1983. Biology and survival of the immature Puffin *Fratercula arctica. Ibis* 125: 56–73.

Harrison, C. J. O. 1969. Helpers at the nest in Australian passerine birds. *Emu* 69: 30–40.

Hartley, P. H. T. 1953. An ecological study of the feeding habits of the English titmice. *J. Anim. Ecol.* 22: 261–88.

Heatwole, H. F. 1965. Some aspects of the association of cattle egrets with cattle. *Anim. Behav.* 13: 79–83.

Heatwole, H. F. and Lowman, M. 1986. *Dieback: death of an Australian landscape.* Sydney: Reed.

Heinsohn, R. G. 1987. Age-dependent vigilance in winter congregations of co-operatively breeding white-winged choughs *Corcorax melanorhamphus. Behav. Ecol. Sociobiol.* 20: 303–6.

Heppelston, P. B. 1972. The comparative breeding ecology of Oystercatchers (*Haemotopus ostralegus* L) in inland and coastal habitats. *J. Anim. Ecol.* 41: 23–51.

Hermes, N. 1981. Mixed-species flocks in a dry sclerophyll forest in autumn and winter. *Corella* 5: 41–5.

Herrera, C. M. 1980. Evolución estacional de las communidades de passeriformes en dos encinares de Andalucía occidental. *Ardeola* 25: 143–80.

Higgs, A. J. 1981. Island biogeography theory and nature reserve design. *J. Biogeogr.* 8: 117–24.

Hindmarsh, A. M. 1986. The functional significance of vocal mimicry in song. *Behaviour* 99: 87–100.

Hindwood, K. A. 1956. The migration of the White-naped and Yellow-faced Honeyeaters. *Emu* 56: 521–25.

Hodgson, A. 1975. Some aspects of the ecology of the Fairy Penguin *Eudyptula minor novaehollandiae* (Forster 1781) in southern Tasmania. Ph.D. thesis, Univ. Tasmania, Hobart.

Hogan-Warburg, A. J. 1966. Social behaviour of the Ruff *Philomachus pugnax* (L.). *Ardea* 54: 109–229.

Hogstad, O. 1976. Sexual dimorphism and divergence in winter foraging behaviour of Three-toed Woodpecker *Picoides tridactylus*. *Ibis* 118: 41–50.

Hollands, D. L. G. 1979. The Letter-winged Kite irruption of 1976-77. *Aust. Bird Watcher* 8: 1–18.

Holyoak, D. T. and Thibault, J-C. 1978. Notes on the phylogeny, distribution and ecology of frugivorous pigeons in Polynesia. *Emu* 78: 201–6.

Hopper, S. D. and Burbidge, A. H. 1986. Speciation of bird-pollinated plants in south-western Australia. In *The dynamic partnership: Birds and plants in southern Australia,* pp. 20–31. *See* Ford 1986a.

Hopper, S. D. and Moran, G. F. 1981. Bird pollination and the mating system of *Eucalyptus stoatei*. *Aust. J. Bot.* 29: 625–38.

Howard, R. D. 1974. The influence of sexual selection and interspecific competition on Mockingbird song (*Mimus polyglottus*). *Evolution* 28: 428–38.

Howarth, D. M. and Grant, T. R. 1982. Species diversity of waterbirds on Lake Illawarra, Coomaditchy Lagoon and the Cordeaux Storage dams, New South Wales. *Emu* 82: 106–9.

Howe, R. W. 1984. Local dynamics of bird assemblages in small forest habitat islands in Australia and North America. *Ecology* 65: 1585–1601.

Howe, R. W., Howe, T. D. and Ford, H. A. 1981. Bird distributions on small rainforest remnants in New South Wales. *Aust. Wildl. Res.* 8: 637–51.

Huddy, L. 1979. Social behaviour and feeding ecology of Scarlet Robins *Petroica multicolor*. Hons thesis, Department of Zoology, Univ. New England, Armidale.

Hughes, R. A. 1985. Notes on the effects of El Niño on the seabirds of the Mollendo district, south west Peru, in 1983. *Ibis* 127: 385–88.

Högstedt G. 1980. Prediction and test of the effects of interspecific competition. *Nature* 283: 64–6.

Jackson, S. W. 1919. Haunts of the Letter-winged Kite *Elanus scriptus* Gould. *Emu* 18: 160–72.

Jaeger, M. M., Bruggers, R. L., Johns, B. E. and Erickson, W. A. 1986. Evidence of itinerant breeding of the Red-billed Quelea *Quelea quelea* in the Ethiopian Rift Valley. *Ibis* 128: 469–82.

Jaensch, R. P. 1984. Censusing Waterbirds using south-west wetlands. In *Methods of censusing birds in Australia,* ed. S. J. J. F. Davies, pp. 34–42. RAOU Report No. 7.

Jenni, D. A. and Collier, G. 1972. Polyandry in the American Jacana *Jacana spinosa*. *Auk* 89: 743–65.

Johnston, R. F. and Selander, R. K. 1971. Evolution in the House Sparrow. II. Adaptive differentiation in North American populations. *Evolution* 25: 1–28.

Jones, D. N. 1981. Temporal changes in the suburban avifauna of an inland city. *Aust. Wildl. Res.* 8: 109–19.

———. 1983. The suburban bird community of Townsville, a tropical city. *Emu* 83: 12–18.

———. 1987. Feeding ecology of the Cockatiel *Nymphicus hollandicus* in a grain-growing area. *Aust. Wild. Res.* 14: 105–15.

———. 1988. Selection of incubation mound sites by the Australian Brush-turkey (*Alectura lathami*). *Ibis* 130: 251–60.

Joseph, L. 1986. The decline and present status of the Black-eared Miner in South Australia. *S. Aust. Orn.* 30: 5-12.

Järvinen, O. and Ulfstrand, S. 1980. Species turnover of a continental bird fauna: Northern Europe, 1850–1970. *Oecologia* 46: 186–95.

Kavanagh, R. P., Shields, J. M., Recher, H. F. and Rohan-Jones, W. G. 1985. Bird populations of a logged and unlogged forest mosaic in the Eden woodchip area. In *Birds of eucalypt forest and woodland: Ecology, conservation, management,* pp. 273–81. *See* Bell 1985b.

Keast, J. A. 1958a. Speciation in the genus *Amytornis* Stejneger (Passeres: Muscicapidae, Malurinae) in Australia. *Aust. J. Ecol.* 6: 33–52.

———. 1958b. The relationship between seasonal movements and the development of geographic variation in the Australian chats (*Ephthianura* Gould and *Ashbyia* North (Passeres: Muscicapidae, Malurinae)). *Aust. Zool.* 6: 53–68.

———. 1961. Bird speciation on the Australian continent. *Bull. Museum Comp. Zool. Harvard* 123: 307–495.

———. 1968a. Competitive interactions and the evolution of ecological niches as illustrated by the honeyeater genus *Melithreptus* (Meliphagidae). *Evolution* 22: 762–84.

———. 1968b. Seasonal movements in the Australian honeyeaters (Meliphagidae) and their ecological significance. *Emu* 67: 159–210.

———. 1985. Bird community structure in southern forests and northern woodlands: a comparison. In *Birds of eucalypt forests and woodlands: Ecology, conservation, management,* pp. 97–116. *See* Bell 1985b.

Keast, J. A., Recher, H. F., Ford, H. A. and Saunders, D. 1985. *Birds of eucalypt forests and woodlands: ecology, conservation, management.* Sydney: RAOU and Surrey Beatty.

Keighery, G. J. 1982. Bird-pollinated plants in Western Australia. In *Pollination and Evolution,* eds. J. A. Armstrong, J. M. Powell and A. J. Richards, pp. 77–90. Sydney: Roy. Bot. Gardens.

Kenyon, R. F. 1972. Polygyny among Superb Lyrebirds. *Emu* 72: 70–76.

Kikkawa, J. 1962. Wintering Silvereyes at bird tables in the Dunedin area. *Notornis* 9: 284–91.

———. 1977. Ecological paradoxes. *Aust. J. Ecol.* 2: 121–36.

———. 1980. Winter survival in relation to dominance classes among Silvereyes *Zosterops lateralis chlorocephala* of Heron Island, Great Barrier Reef. *Ibis* 122: 437–46.

Kikkawa, J. and Pearse, K. 1969. Geographical distribution of land birds in Australia — a numerical analysis. *Aust. J. Zool.* 17: 821–40.

Kikkawa, J. and Wilson, J. M. 1983. Breeding and dominance among the Heron Island Silvereye *Zosterops lateralis chlorocephala. Emu* 83: 181–98.

King, B. R. 1980. Social organization and behaviour of the Grey-crowned Babbler *Pomatostomus temporalis. Emu* 80: 59–76.

King, W. B. 1977. Endangered birds of the world and current efforts toward managing them. In *Endangered Birds,* pp. 9–17. *See* Flack 1977.

———. 1981. *Endangered Birds of the World. The ICBP Bird Red Data Book.* Washington, D. C.: Smithsonian Institution Press.

Kitchener, D. J., 1982. Predictors of vertebrate species richness in nature reserves in the Western Australian wheatbelt. *Aust. Wildl. Res.* 9: 1–7.

Kitchener, D. J., Dell, J. and Muir, B. G. 1982. Birds in Western Australian wheatbelt reserves — implications for conservation. *Biol. Conserv.* 22: 127–63.

Kluijver, H. N. 1966. Regulation of a bird population. *Ostrich* suppl. 6: 389.

Koenig, W. 1981. Reproductive success, group size and the evolution of cooperative breeding in the Acorn Woodpecker. *Am. Nat.* 117: 421–43.

———. 1986. Geographical ecology of clutch size variation in North American woodpeckers. *Condor* 88: 499–504.

Krebs, C. 1972. *Ecology, the experimental analysis of distribution and abundance.* New York: Harper and Row.

Krebs, J. R. 1978. Optimal foraging: decision rules for predators. In *Behavioural Ecology: an evolutionary approach,* pp. 23–63, *See* Davies 1978.

Lack, D. 1954. *The natural regulation of animal numbers.* Oxford: Oxford Univ. Press.

———. 1965. *The Life of the Robin.* London: Collins.

———. 1966. *Population studies of birds.* Oxford: Oxford Univ. Press.

———. 1976. *Island Biology.* Oxford: Blackwell.

———. 1983. *Darwins Finches.* Cambridge Univ. Press: Cambridge.

Landres, P. B. and MacMahon, J. A. 1983. Community organization of arboreal birds in some oak woodlands of western North America. *Ecol. Monogr.* 58: 183–208.

Lane, B. A. 1987. *Shorebirds of Australia.* Melbourne: Nelson.

Lane, B. A. and Jessop, A. E. 1985. Tracking of migrating waders in north-western Australia using meteorological radar. *Stilt* 6: 17–28.

Lane, S. G. 1972. Tasmanian type Silvereyes in New South Wales. *Aust. Bird Bander* 10: 33–4.

Lange, R. T. 1982. Australian Tertiary vegetation: evidence and interpretation. In *A History of Australasian Vegetation,* ed. J. M. B. Smith, pp. 44–89. Sydney: McGraw-Hill.

Larson, D. L. and Bock, C. E. 1986. Eruptions of some North American boreal seed-eating birds 1901–1980. *Ibis* 128: 137–40.

Lavery, H. J. 1970. Saltwater habitat as a factor in the ecology of some Australian waterfowl, particularly Grey Teal (*Anas gibberifrons*). Ph.D. thesis, Univ. of Queensland.

Lea, A. M. and Gray, J. T. 1935a. The food of Australian birds. Part I. *Emu* 34: 275–92.

———. 1935b. The food of Australian birds. Part II. *Emu* 35: 63–97.

———. 1935c. The food of Australian birds. Part III. *Emu* 35: 145–78.

———. 1936a. The food of Australian birds. Part IV. *Emu* 36: 251–80.

———. 1936b. The food of Australian birds. Part V. *Emu* 36: 335–47.

Liddy, J. 1966. Autumnal migration of Yellow-faced Honeyeaters. *Emu* 66: 87–103.

Lill, A. 1974. Sexual behaviour of the lek-forming White-bearded Manakin (*Manacus manacus trinitatis*). *Z. Tierpsychol.* 36: 1–36.

———. 1979. An assessment of male parental investment and pair bonding in the polygamous Superb Lyrebird. *Auk* 96: 489–98.

———. 1986. Time-energy budgets during reproduction and the evolution of single parenting in the Superb Lyrebird. *Aust. J. Zool.* 34: 351–71.

Lofts, B. and Murton, R. K. 1968. Photoperiodic and physiological adaptations regulating avian breeding cycles and their ecological significance. *J. Zool.* 155: 327–94.

Long, J. L. 1981. *Introduced Birds of the World.* Sydney: Reed.

———. 1984. The diets of three species of parrots in the south of Western Australia. *Aust. Wildl. Res.* 11: 357–71.

Longmore, N. W. and Boles, W. E. 1983. Description and systematics of the Eungella Honeyeater from central Queensland. *Emu* 83: 59–65.

Lowe, 1982. Feeding behaviour and diet of Royal Spoonbills *Platalea regia* in Westernport Bay, Victoria. *Emu* 82: 163–68.

———. 1983. Feeding behaviour and diet of the White-faced Heron *Ardea novaehollandiae* in Westernport Bay, Victoria. *Corella* 7: 101–8.

Lowman, M. D. 1982. Seasonal variation in insect abundance among three Australian rainforests, with particular reference to phytophagous types. *Aust. J. Ecol.* 7: 353–63.

Loyn, R. H. 1980. Bird populations in a mixed eucalypt forest used for production of wood. *Emu* 80: 145–56.

———. 1985. Ecology, distribution and density of birds in Victorian forests. In *Birds of eucalypt forests and woodlands: ecology, conservation, management,* pp. 323–31. *See* Bell 1985b.

———. 1987. Effects of patch area and habitat on bird abundances, species numbers and tree health in fragmented Victorian forests. In *Nature Conservation: the role of remnants of native vegetation,* pp. 65–77. *See* Saunders *et al.* 1987.

Loyn, R. H., Runnals, R. G., Forward, G. Y. and Tyers, J. 1983. Territorial Bell Miners and other birds affecting populations of insect prey. *Science* 221: 1411–13.

MacArthur, R. H. 1958. Population ecology of some warblers of north-eastern coniferous forests. *Ecology* 39: 599–619.

MacArthur, R. H. and Levins, R. 1967. The limiting similarity, convergence and divergence of coexisting species. *Am. Nat.* 101: 377–85.

MacArthur, R. H. and MacArthur, J. W. 1961. On bird species diversity. *Ecology* 42: 594–98.

MacArthur, R. H. and Wilson, E. O. 1967. *The theory of island biogeography.* Princeton, N. J.: Princeton Univ. Press.

Maclean, G. L. 1976. Rainfall and avian breeding seasons in north-western New South Wales in spring and summer 1974-75. *Emu* 76: 139–142.

Marchant, S. 1972. Evolution of the genus *Chrysococcyx. Ibis* 114: 219–33.

———. 1974. Analysis of nest-records from the Willie Wagtail. *Emu* 74: 149–60.

———. 1980. Incubation and nestling periods of some Australian birds. *Corella* 4: 30–2.

———. 1982. The sedentary nature of passerine birds in woodland at Moruya, New South Wales. *Corella* 6: 87–9.

———. 1984. Nest-records of the Eastern Yellow Robin *Eopsaltria australis. Emu* 84: 167–74.

———. 1985. Breeding of the Eastern Yellow Robin. In *Birds of eucalypt forests and woodlands: Ecology, conservation, management,* pp. 231–240. *See* Bell 1985b.

Marchant, S. and Fullagar, P. 1983. Nest records of the Welcome Swallow. *Emu* 83: 66–74.

Marchant, S. and Whittington, P. 1986. 1984–85 CBC Index Report. *BTO News* 146: 7–10.

Marshall, A. J. 1954. *Bower-birds.* Oxford: Clarendon.

Marshall, A. J. and Serventy, D. 1958. The internal rhythm of reproduction in xerophilous birds under conditions of illumination and darkness. *J. Exp. Biol.* 35: 666–70.

Martindale, J. 1986. The Freckled Duck — an RAOU Conservation Statement. RAOU Report No. 22, Melbourne.

Mason, C. F. 1976. Breeding biology of the *Sylvia* warblers. *Bird Study* 23: 213-32.

Matheson, W. E. 1978. A further irruption of native-hens in 1975. *S. Aust. Orn.* 27: 270–73.

Matthiessen, J. N. and Springett, B. P. 1973. The food of the Silvereye *Zosterops gouldi* (Aves: Zosteropidae), in relation to its role as a vector of the granulosis virus of the potato moth *Phthorimaea operculata* (Lepidoptera: Gelechiidae). *Aust. J. Zool.* 21: 533–40.

May, R. M. 1972. Will a large complex system be stable? *Nature* 238: 413–14.

Mayr, E. 1942. *Systematics and the origin of species*. New York: Columbia Univ. Press.

———. 1944. Timor and the colonization of Australia by birds. *Emu* 44: 113–30.

McClure, H. E. 1974. *Migration and survival of birds of Asia*. Applied Scientific Research Corporation of Thailand, Bangkok.

McFarland, D. C. 1985. Flowering biology and phenology of *Banksia integrifolia* and *B. spinulosa* (Proteaceae) in New England National Park, N.S.W. *Aust. J. Bot.* 33: 705–14.

———. 1986a. The organization of a honeyeater community in an unpredictable environment. *Aust. J. Ecol.* 11: 107–20.

———. 1986b. Determinants of feeding territory size in the New Holland Honeyeater *Phylidonyris novaehollandiae*. *Emu* 86: 180–85.

McFarland, D. C. and Ford, H. A. 1987. Aspects of population biology of the Eastern Spinebill *Acanthorhynchus tenuirostris* (Meliphagidae) in New England National Park, New South Wales. *Corella* 11: 52–8.

McGilp, J. N. 1932. Heat in the interior of South Australia and in central Australia. Holocaust of Bird-life. *S. Aust. Orn.* 11: 158–63.

McGilp, J. and Morgan, A. M. 1931. The nesting of the Banded Stilt. *S. Aust. Orn.* 11: 37–53.

McKeown, K. C. 1936. The food of Australian birds with reference to the protective adaptations in insects. *Emu* 36: 21–30.

———. 1944. Notes on the food of Australian birds. *Emu* 43: 188–91, 259–69.

Meredith, C. W. 1984. The Ground Parrot *Pezoporus wallicus* Kerr. RAOU Conservation Statement No. 1, Melbourne.

Meredith, C. W., Gilmore, A. M. and Isles, A. C. 1984. The Ground Parrot (*Pezoporus wallicus* Kerr) in south-eastern Australia: a fire-adapted species? *Aust. J. Ecol.* 9: 367–80.

Merton, D. V., Morris, R. B. and Atkinson, I. A. E. 1984. Lek behaviour in a parrot: the Kakapo *Strigops labroptilus* of New Zealand. *Ibis* 126: 272–83.

Middleton, A. L. A. 1970a. Foods and feeding habits of the European Goldfinch near Melbourne. *Emu* 70: 12–16.

———. 1970b. The breeding biology of the Goldfinch in south-eastern Australia. *Emu* 70: 159–67.

Milewski, A. V. 1986. A comparison of bird-plant relationships in southern Australia and southern Africa. In *The dynamic partnership: Birds and plants in southern Australia,* pp. 111–18. *See* Ford 1986a.

Milledge, D. R. and Recher, H. F. 1985. A comparison of forest bird communities on the New South Wales south and mid-north coasts. In *Birds of eucalypt forests and woodlands: ecology, conservation, management,* pp. 47–52. *See* Bell 1985b.

Miller, B. and Mullette, K. J. 1985. Rehabilitation of an endangered Australian bird: the Lord Howe Island Woodhen *Tricholimnas sylvestris* (Sclater). *Biol. Conserv.* 34: 55–95.

Mills, J. A., Lavers, R. B. and Lee, W. G. 1988. The post-Pleistocene decline of takahe (*Notornis mantelli*): a reply. *J. Roy. Soc. N.Z.* 18: 112–18.

Mills, J. A. and Williams, G. R. 1979. The status of endangered New Zealand birds. In *The status of endangered Australasian wildlife,* pp. 147–68. *See* Cowling 1979.

Minot, E. O. 1981. Effects of interspecific competition for food in breeding blue and great tits. *J. Anim. Ecol.* 50: 375–85.

Minton, C. D. T. and Lane, B. A. 1984. Counting and banding waders in Australia and New Zealand. In *Methods of censusing birds in Australia,* pp. 43–54. *See* Jaensch 1984.

Moreau, R. E. 1944. Clutch size: a comparative study, with special reference to African birds. *Ibis* 86: 286–347.

Morgan, R. 1982. The breeding biology of the Nightingale *Luscinia megarhynchos* in Britain. *Bird Study* 29: 67–72.

Morse, D. G. 1970. Ecological aspects of some mixed-species foraging flocks of birds. *Ecol. Monogr.* 40: 119–67.

———. 1978. Structure and foraging patterns of flocks of tits and associated species in an English woodland during the winter. *Ibis* 120: 295–310.

Moss, D. 1978. Diversity of woodland song-bird populations. *J. Anim. Ecol.* 47: 521–27.

Moynihan, M. 1962. The organization and probable evolution of some mixed-species flocks of Neotropical birds. *Smithson. misc. Coll.* 143: 1–140.

———. 1968. Social mimicry: character convergence versus character displacement. *Evolution* 22: 315–31.

Murray, B. G. 1981. The origins of adaptive interspecific territorialism. *Biol. Rev.* 56: 1–22.

Murton, R. K. and Westwood, N. J. 1977. *Avian breeding cycles.* Oxford: Clarendon.

Myers, J. P., Conners, P. G. and Pitelka, F. A. 1979. Territory size in wintering sanderlings: the effect of prey abundance and intruder density. *Auk* 96: 551–61.

Møller, A. P. 1986. Mating systems among European passerines: a review. *Ibis* 128: 234–50.

Naarding, J. A. 1981. Population estimates of the Short-tailed Shearwater *Puffinus tenuirostris* in Tasmania. *Tas. Nat.* 66: 3–4.

Nelson, J. B. 1964. Factors influencing clutch-size and chick growth in the North Atlantic Gannet *Sula bassana. Ibis* 106: 63–77.

———. 1978. *The Sulidae: Gannets and Boobies.* Oxford Univ. Press.

Nelson, J. B. and Powell, D. 1986. The breeding ecology of Abbott's Booby *Sula abbotti. Emu* 86: 33–46.

Newton, I. 1964. The breeding biology of the Chaffinch. *Bird Study* 11: 47–68.

———. 1972. *Finches.* London: Collins.

———. 1979. *Population Ecology of Raptors.* Calton, Staffordshire: Poyser.

Nias, R. C. 1984. Territory quality and group size in the Superb Fairy-wren *Malurus cyaneus. Emu* 84: 178–80.

———. 1986. Nest-site characteristics and reproductive success in the Superb Fairy-wren. *Emu* 86: 139–44.

———. 1987. Co-operative breeding in the Superb Fairy-wren (*Malurus cyaneus*). Ph.D. thesis, University of New England, Armidale, New South Wales.

Nice, M. M. 1957. Nesting success in altricial birds. *Auk* 74: 305–21.

Nicholson, A. J. 1933. The balance of animal populations. *J. Anim. Ecol.* 2: 132–78.

Niethammer, G. 1970. Clutch sizes of introduced European passeriformes in New Zealand. *Notornis* 17: 214–22.

Nix, H. A. 1976. Environmental control of breeding, post-breeding dispersal and migration of birds in the Australian region. *16th Int. Orn. Congr.* (Canberra.) pp. 272–305.

Nolan, V. 1963. Reproductive success in a deciduous scrub habitat. *Ecology* 44: 305–13.

Norberg, U. M. 1979. Morphology of the wings, legs and tail of three coniferous forest tits, the Goldcrest, and the Treecreeper in relation to locomotor pattern and feeding station selection. *Phil. Trans. R. Soc.* (Lond.) 287(B): 131–65.

Noske, R. A. 1983. Comparative behaviour and ecology of some Australian bark-foraging birds. Ph.D. thesis, Univ. of New England, Armidale, New South Wales.

———. 1985. Habitat use by three bark-foragers of eucalypt forests. In *Birds of eucalypt forests and woodlands: ecology, conservation, management,* pp. 193–204. *See* Bell 1985b.

———. 1986. Intersexual niche segregation among three bark-foraging birds of eucalypt forests. *Aust. J. Ecol.* 11: 255–67.

O'Connor, R. J. 1985. Long-term monitoring of British bird populations. *Ornis Fennica* 62: 73–9.

O'Connor, R. J. and Shrubb, M. 1986. *Farming and Birds.* Cambridge Univ. Press.

Odum, E. P. 1959. *Fundamentals of Ecology.* Philadelphia: Saunders.

Old, K. M., Kile, G. A. and Ohmart, C. P. 1979. *Eucalypt dieback in forests and woodlands.* Melbourne: CSIRO.

Olsen, P. 1986. From the President. *ARA News* 7: 61.

Olson, S. L. and Feduccia, A. 1980. Relationships and evolution of flamingos (Aves: Phoenicopteridae). *Smithsonian Contrib. Zool.* 316: 1–73.

Olson, S. L. and James, H. F. 1982. Fossil birds from the Hawaiian islands: evidence for wholesale extinction by man before western contact. *Science* 217: 633–35.

Orell, M. and Ojanen, M. 1983. Effect of habitat, date of laying and density on clutch size of the Great Tit in northern Finland. *Holarctic Ecol.* 6: 413–23.

Orians, G. H. 1969. On the evolution of mating systems in birds and mammals. *Am. Nat.* 103: 589–603.

———. 1980. *Some adaptations of marsh-nesting Blackbirds.* Princeton, New Jersey: Princeton Univ. Press.

Otvos, I. S. 1979. The effects of insectivorous bird activities in forest ecosystems: an evaluation. In *The role of insectivorous birds in forest ecosystems.* eds. J. G. Dickson, R. N. Connor, R. R. Fleet, J. A. Jackson and J. C. Kroll, pp. 341–374. New York: Academic Press.

Parmelee, D. F. and Payne, R. B. 1973. On multiple broods and the breeding strategy of Arctic Sanderlings. *Ibis* 115: 219–26.

Paton, D. C. 1979. The behaviour and feeding ecology of the New Holland honeyeater *Phylidonyris novaehollandiae* in Victoria. Ph.D. thesis, Department of Zoology, Monash University.

———. 1980. The importance of manna, honeydew and lerp in the diet of honeyeaters. *Emu* 80: 213–26.

———. 1982. The diet of the New Holland honeyeater *Phylidonyris novaehollandiae. Aust. J. Ecol.* 7: 279–98.

———. 1985a. Food supply, population structure and behaviour of New Holland Honeyeaters *Phylidonyris novaehollandiae* in woodland near Horsham, Victoria. In *Birds of eucalypt forest and woodland: ecology, conservation and management,* pp. 219–30. *See* Bell 1985b.

———. 1985b. Do New Holland Honeyeaters *Phylidonyris novaehollandiae* breed regularly in spring and autumn? *Emu* 85: 130–33.

———. 1986a. Honeyeaters and their plants in south-eastern Australia. In *The dynamic partnership: birds and plants in southern Australia,* pp. 9–19. *See* Ford 1986a.

———. 1986b. Evolution of bird pollination in Australia. In *The dynamic partnership: birds and plants in Southern Australia,* pp. 32–41. *See* Ford 1986a.

Paton, D. C. and Ford, H. A. 1977. Pollination by birds of native plants in South Australia. *Emu* 76: 73-85.

———. 1983. The influence of plant characteristics and honeyeater size on levels of pollination in Australian plants. In *Handbook of Experimental Pollination Biology,* eds. C. E. Jones and R. J. Little, pp. 235–248, Princeton, New Jersey: Van Nostrand-Reinhold.

Paton, D. C. and Turner, V. 1985. Pollination of *Banksia ericifolia* Smith: Birds, mammals and insects as pollen vectors. *Aust. J. Bot.* 33: 271–86.

Pearson, D. L. 1977. A pantropical comparison of bird community structure on six lowland forest sites. *Condor* 79: 232–42.

Perrins, C. M. 1979. *British Tits.* Glasgow: Collins.

Perrins, C. M. and Birkhead, T. R. 1983. *Avian Ecology.* Glasgow: Blackie.

Pianka, E. R. 1974. *Evolutionary Ecology.* New York: Harper and Row.

Pimm, S. L. 1984. The complexity and stability of ecosystems. *Nature* 307: 321–26.

Pitelka, F. A. 1941. Distribution of birds in relation to major biotic communities. *Am. Midl. Nat.* 25: 113–37.

Pratt, T. K. 1984. Examples of tropical frugivores defending fruit-bearing plants. *Condor* 86: 123–29.

Pruett-Jones, M. A. and Pruett-Jones, S. G. 1982. Spacing and distribution of bowers in Macgregor's Bowerbird (*Amblyornis macgregoriae*). *Behav. Ecol. Sociobiol.* 11: 25–32.

———. 1985. Food-caching in the tropical frugivore, Macgregor's Bowerbird (*Amblyornis macgregoriae*). *Auk* 102: 334–41.

Pruett-Jones, S. G. and Pruett-Jones, M. A. 1986. Altitudinal distribution and seasonal activity patterns of birds of paradise. *Natl. Geog. Res.* 2: 87–105.

———. 1988a. The use of court objects by Lawes' Parotia. *Condor* 90: 538–45.

———. 1988b. A promiscuous mating system in the Blue Bird of Paradise *Paradisaea rudolphi. Ibis* 130: 373–77.

Purchase, D. 1985. Bird-banding and the migration of Yellow-faced and White-naped Honeyeaters through the Australian Capital Territory. *Corella* 9: 59–62.

Pyke, G. H. 1980. The foraging behaviour of Australian Honeyeaters: a review and some comparisons with hummingbirds. *Aust. J. Ecol.* 5: 343–69.

———. 1982. Evolution of inflorescence size and height in Waratahs (*Telopea speciosissima*): the difficulties of interpreting correlations between plant traits and fruit set. In *Pollination and evolution,* pp. 91–4. *See* Keighery 1982.

———. 1983. Seasonal pattern of abundance of honeyeaters and their resources in heathland areas near Sydney. *Aust. J. Ecol.* 8: 217–33.

Raikow, R. J. 1976. The origin and evolution of the Hawaiian Honeycreepers (Drepanididae). *Living Bird* 15: 95–117.

Readshaw, J. L. 1968. The distribution, abundance and seasonal movements of pied currawong *Strepera graculina* (Shaw), an important bird predator of Phasmatidae in eastern Australia. *Aust. J. Zool.* 16: 37–47.

Recher, H. F. 1969. Bird species diversity and habitat diversity in Australia and North America. *Am. Nat.* 103: 75–80.

———. 1971. Sharing of habitat by three congeneric honeyeaters. *Emu* 71: 147–52.

———. 1977. Ecology of co-existing White-cheeked and New Holland Honeyeaters. *Emu* 77: 136–42.

———. 1981. Nectar feeding and its evolution in Australia. In *Ecological Biogeography in Australia,* pp. 1639–47, ed. J. A. Keast. The Hague: Junk.

———. 1984. Use of bird census procedures in Australia: A review. In *Methods of censusing birds in Australia,* pp. 3–14, *See* Jaensch 1984.

———. 1985. Synthesis: a model of forest and woodland bird communities. In *Birds of eucalypt forests and woodlands: ecology, conservation, management,* pp. 129–135. *See* Bell 1985b.

Recher, H. F. and Abbott, I. J. 1970. Some differences in the use of habitat by White-eared and White-cheeked Honeyeaters. *Emu* 70: 117–25.

Recher, H. F., Allen, D. and Gowing, G. 1985. The impact of wildfire on birds in an intensively logged forest. In *Birds of eucalypt forests and woodlands: ecology, conservation, management,* pp. 283–290. *See* Bell 1985b.

Recher, H. F., Davis, W. E. and Holmes, R. T. 1987. Ecology of Brown and Striated Thornbills in forests of south-eastern New South Wales, with comments on forest management. *Emu* 87: 1–13.

Recher, H. F. and Holmes, R. T. 1985. Foraging ecology and seasonal patterns of abundance in a forest avifauna. In *Birds of eucalypt forests and woodlands: ecology, conservation, management,* pp. 79–96. *See* Bell 1986b.

Recher, H. F., Holmes, R. T. Davis, W. E. and Morton, S. 1983. Foraging behaviour of Australian herons. *Colonial Waterbirds* 6: 1–10.

Recher, H. F., Holmes, R. T., Schultz, M., Shields, J. and Kavanagh, R. 1985. Foraging patterns of birds in eucalypt forest and woodland on the tablelands of south-eastern Australia. *Aust. J. Ecol.* 10: 399–419.

Recher, H. F. and Recher, J. T. 1972. The foraging behaviour of the Reef Heron. *Emu* 72: 85–90.

Recher, H. F., Shields, J., Kavanagh, R. and Webb, G. 1987a. Retaining remnant mature forest for native conservation at Eden, New South Wales: a review of theory and practice. In *Nature Conservation: the role of remnants of native vegetation,* pp. 177–94. *See* Saunders *et al.* 1987.

Recher, H. F. and Shultz, M. 1983. Observations on the breeding of White-browed Woodswallows. *Corella* 7: 1–6.

Reilly, P. N. 1970. Nesting of the Superb Lyrebird in Sherbrooke Forest, Victoria. *Emu* 70: 73–78.

Reilly, P. N. and Cullen, J. M. 1979. The Little Penguin *Eudyptula minor* in Victoria 1. Mortality of adults. *Emu* 79: 97–102.

———. 1981. The Little Penguin *Eudyptula minor* in Victoria II. Breeding. *Emu* 81: 1–19.

———. 1982. The Little Penguin *Eudyptula minor* in Victoria III. Dispersal of chicks and survival after banding. *Emu* 82: 127–42.

Rich, P. V. 1975. Changing continental arrangements and the origin of Australia's non-passeriform continental avifauna. *Emu* 75: 97–112.

Rich, P. V. and van Tets, J. 1982. Fossil birds of Australia and New Guinea: Their biogeographic, phylogenetic and biostratigraphic input. In *The fossil vertebrate record of Australia,* eds. P. V. Rich and E. M. Thompson, pp. 235–384. Melbourne: Monash Univ.

Rich, P. V. and Baird, R. F. 1986. History of the Australian avifauna. *Current Ornithology* 4: 97–139.

Richardson, K. C. and Wooller, R. D. 1986. The structures of the gastro-intestinal tracts of honeyeaters and other small birds in relation to their diets. *Aust. J. Zool.* 34: 119–24.

Richardson, K. C. and Wooller, R. D. 1989. The alimentary tract of a specialist frugivore: the Mistletoebird *Dicaeum hirundinaceum* in relation to its diet. *Aust. J. Zool.,* in press.

Richdale, L. E. 1957. *A population study of penguins.* Oxford: Oxford Univ. Press.

Ricklefs, R. E. 1980a. *Ecology.* Walton on Thames: Nelson.

———. 1980b. Geographical variation in clutch size among passerine birds: Ashmole's hypothesis. *Auk* 97: 38–49.

Robertson, J. S. and Woodall, P. F. 1987. Survival of Brown Honeyeaters in south-east Queensland. *Emu* 87: 137–42.

Robinson, F. N. 1974. The function of vocal mimicry in some avian displays. *Emu* 74: 9–10.

Rooke, I. J. 1979. The social behaviour of the honeyeater *Phylidonyris novaehollandiae.* Ph.D. thesis, Univ. of Western Australia. RAOU Microfiche No. 3: Melbourne.

———. 1984. The Silvereye (*Zosterops lateralis:* Aves: Zosteropidae): a review. *J. Roy. Soc. W.A.* 66: 163–169.

Rose, A. B. 1973. Food of some Australian birds. *Emu* 73: 177–83.

Rowley, I. 1962. 'Rodent run' distraction display by a passerine, the Superb Blue Wren *Malurus cyaneus* L. *Behaviour* 19: 170–76.

———. 1965. The life history of the Superb Blue Wren *Malurus cyaneus. Emu* 64: 251–97.

———. 1968. Communal species of Australian birds. *Bonn. Zool. Beitr.* 19: 362–70.

———. 1973a. The comparative ecology of Australian corvids. II. Social organization and behaviour. *CSIRO Wildl. Res.* 18: 25–65.

———. 1973b. The comparative ecology of Australian corvids. IV. Nesting and the rearing of young to independence. *CSIRO Wildl. Res.* 18: 91–129.

———. 1974. *Bird Life.* Sydney: Collins.

———. 1976. Co-operative breeding in Australian birds. *Proc. 16th Int. Orn. Congr.* (Canberra). 657–66.

———. 1978. Communal activities among White-winged Choughs, *Corcorax melanorhamphus. Ibis* 120: 178–97.

———. 1981. The communal way of life in the Splendid Wren *Malurus splendens. Z. Tierpsychol.* 55: 228–67.

———. 1983. Mortality and dispersal of juvenile Galahs *Cacatua roseicapilla,* in the Western Australian wheatbelt. *Aust. Wildl. Res.* 10: 329–42.

Ruiz, X., Jover, L., Worente, G. A., Sanchez-Reyes, A. F. and Febrian, M. I. 1988. Song Thrush *Turdus philomelos* wintering in Spain as biological indicators of the Chernobyl accident. *Ornis. Scand.* 19: 63–6.

Russell, E. and Rowley, I. 1988. Helper contributions to reproductive success in the Splendid Fairy-Wren (*Malurus splendens*). *Behav. Ecol. Sociobiol.* 22. 131–40.

Ryan, J. V. 1954. Further notes on the Rufous Songlark. *Emu* 54: 128–31.

Sagar, P. 1986. Wader counts in New Zealand. *Stilt* 9: 32–34.

Salomonsen, F. 1985. Flowerpecker. In *A dictionary of birds*, pp. 228–30. *See* Frith 1985.

Sara, G. J. 1982. A Grey Wagtail in southern Australia. *S. Aust. Orn.* 28: 220.

Sargent, O. H. 1928. Reactions between birds and plants. *Emu* 27: 185–92.

Saunders, D. A. 1974. Subspeciation in the White-tailed Black Cockatoo *Calyptorhynchus baudinii,* in Western Australia. *Aust. Wildl. Res.* 1: 55–69.

———. 1977. Red-tailed Black Cockatoo breeding twice a year in the south-west of Western Australia. *Emu* 77: 107–10.

———. 1979. Distribution and taxonomy of the White-tailed and Yellow-tailed Black Cockatoos *Calyptorhynchus. Emu* 79: 215–27.

———. 1985. Human impact: the response of forest and woodland bird communities. Whither the future? A Synthesis. In *Birds of eucalypt forests and woodlands: ecology, conservation, management,* pp. 355–7. *See* Bell 1985b.

Saunders, D. A., Arnold, G. W., Burbidge, A. A. and Hopkins, A. J. M. 1987. *Nature Conservation: the role of remnants of native vegetation.* Surrey Beatty: Sydney.

Saunders, D. A. and Ingram, J. A. 1987. Factors affecting survival of breeding populations of Carnaby's Cockatoo *Calyptorhynchus funereus latirostris* in remnants of native vegetation. In *Nature Conservation: the role of remnants of native vegetation,* pp. 249–58. *See* Saunders *et al.* 1987.

Saunders, D. A., Rowley, I. and Smith, G. T. 1985. The effects of clearing for agriculture on the distribution of Cockatoos in the south-west of Western Australia. In *Birds of eucalypt forests and woodlands: ecology, conservation, management,* pp. 309–21. *See* Bell 1985b.

Schodde, R. 1979. The status of endangered Papuasian birds. In *The status of endangered Australasian wildlife,* pp. 133–46. *See* Cowling 1979.

Schodde, R., Fullagar, P. and Hermes, N. 1983. *A review of Norfolk Island birds: past and present.* Aust. Nat. Pks and W.S. Spec. Publ. 8.

Schoener, T. W. 1983. Field experiments on interspecific competition. *Am. Nat.* 122: 240–85.

Scott, J. M. and Kepler, C. B. 1985. Distribution and abundance of Hawaiian native birds: a status report. *Bird Conservation* 2: 43–70.

Sedgwick, E. H. 1949. Mixed associations of small birds in the south-west of Western Australia. *Emu* 49: 9–13.

Selander, R. K. 1966. Sexual dimorphism and differential niche utilization in birds. *Condor* 68: 113–51.

———. 1972. Sexual selection and dimorphism in birds. In *Sexual selection and the ascent of man,* ed. B. Campbell, pp. 180-230. Chicago: Aldine Press.

Serventy, D. L. 1953. Some speciation problems in Australian birds. *Emu* 53: 131–45.

———. 1963. Egg-laying timetable of the Slender-billed Shearwater *Puffinus tenuirostris. Proc. 13th Int. Orn. Congr.* 338–43.

———. 1967. Aspects of the population ecology of the Short-tailed Shearwater *Puffinus tenuirostris. Proc. 14th Int. Orn. Congr.* 165–90.

———. 1971. Biology of desert birds. In *Avian Biology. Vol. 1.* eds. D. S. Farner and J. R. King. New York: Academic Press.

Serventy, D. L. and Marshall, A. J. 1957. Breeding periodicity in Western Australian birds: with an account of unseasonal nestings in 1953 and 1955. *Emu* 57: 99–126.

Sharrock, J. T. R. 1976. *The atlas of breeding birds in Britain and Ireland.* England: BTO.

Shields, J. M. and Recher, H. F. 1984. Breeding bird censuses: an evaluation of four methods for use in sclerophyll forest. *Corella* 8: 29–41.

Sibley, C. G. and Ahlquist, J. E. 1985. The phylogeny and classification of the Australo-Papuan passerine birds. *Emu* 85: 1–14.

Sibley, C. G., Williams, G. R. and Ahlquist, J. E. 1982. The relationships of the New Zealand wrens (Acanthisittidae) as indicated by DNA-DNA hybridization. *Notornis* 29: 113–30.

Simberloff, D. 1978. Using island biogeographic distributions to determine if colonization is stochastic. *Am. Nat.* 112: 713–26.

Simberloff, D. and Abele, L. G. 1982. Refuge design and island biogeographic theory: effects of fragmentation. *Am. Nat.* 120: 41–50.

———. 1984. Conservation and obfuscation: subdivision of reserves. *Oikos* 42: 399–401.

Simms, E. 1978. *British Thrushes.* London: Collins.

Singh, G. 1982. Environmental Upheaval: Vegetation of Australasia during the Quaternary. In *A History of Australasian Vegetation,* pp. 90–108. *See* Lange 1982.

Skutch, A. F. 1935. Helpers at the nest. *Auk* 52: 257–73.

———. 1974. *The life of the hummingbird.* London: Octopus.

———. 1987. Helpers at Bird's nests: a worldwide survey of co-operative breeding and related behaviour. Iowa: Iowa Univ. Press.

Smith, A. J. and Robertson, B. I. 1978. Social organization of Bell Miners. *Emu* 78: 169–78.

Smith, G. T. 1977. The effect of environmental change on six rare birds. *Emu* 77: 173–79.

———. 1979a. The Noisy Scrub-bird. In *The status of endangered Australasian wildlife,* pp. 117–21. *See* Cowling 1979.

———. 1979b. The status of Australian parrots. In *The status of endangered Australasian wildlife,* pp. 101–8. *See* Cowling 1979.

———. 1985. The Noisy Scrub-bird *Atrichornis clamosus.* Does its past suggest a future? In *Birds of eucalypt forests and woodlands: ecology, conservation, management,* pp. 301–8. *See* Bell 1985b.

Smith, J. N. M. 1974a. The food searching behaviour of two European thrushes. I. Description and analysis of search paths. *Behaviour* 48: 276–302.

———. 1974b. The food searching behaviour of two European thrushes. II. The adaptiveness of the search patterns. *Behaviour* 49: 1–61.

Smith, J. M. N., Montgomerie, R. D., Taitt, M. J. and Yom-Tov, Y. 1980. A winter feeding experiment on an island song sparrow population. *Oecologia* 47: 164–70.

Smith, M. J. and Brereton, J. Le G. 1976. Annual gonadal and adrenal cycles in the Eastern Rosella *Platycercus eximius* (Psittaciformes: Platycercidae). *Aust. J. Zool.* 24: 541–56.

Smith, P. 1985a. Effects of intensive logging on birds in eucalypt forest near Bega, New South Wales. *Emu* 85: 15–21.

———. 1985b. Woodchip logging and birds near Bega, New South Wales. In *Birds of eucalypt forests and woodlands: ecology, conservation, management,* pp. 259–71. *See* Bell 1985b.

———. 1986. Monitoring the populations and movements of Australian birds: A project proposal. RAOU Report No. 25.

Snow, B. K. and Snow, D. W. 1972. Feeding niches of hummingbirds in a Trinidad valley. *J. Anim. Ecol.* 41: 471–86.

Snow, D. W. 1955. The breeding of Blackbird, Song Thrush and Mistle Thrush in Great Britain. *Bird Study* 2: 169–78.

———. 1958. The breeding of the Blackbird, *Turdus merula* at Oxford. *Ibis* 100: 1–30.

———. 1976. *The web of adaptation.* New York: Quadrangle.

Solonen, T. 1986. Stability of a bird community in southern Finland. *Ornis Scand.* 17: 106–16.

Southwood, T. R. E. 1966. *Ecological Methods.* London: Methuen.

Stokes, T. 1988. A review of the birds of Christmas Island, Indian Ocean. Occasional Paper of Aust. Nat. Pks and Wildl. Serv. 16.

Sutcliffe, S. J. 1986. Changes in the gull populations of SW Wales. *Bird Study* 33: 91–7.

Swainson, G. W. 1970. Co-operative rearing in the Bell Miner. *Emu* 70: 183–88.

Takekawa, J. Y. and Garton, E. O. 1984. How much is an Evening Grosbeak worth? *J. Forestry* 82: 426–28.

Templeton, A. R. 1980. The theory of speciation via the founder principle. *Genetics* 94: 1011–38.

Thomas, D. G. 1980a. Foraging of honeyeaters in an area of Tasmanian sclerophyll forest. *Emu* 80: 55–58.

———. 1980b. The bird community of Tasmanian temperate rainforest. *Ibis* 122: 298-306.

Tiainen, J. 1985. Monitoring bird populations in Finland. *Ornis Fennica* 62: 80–89.

Tidemann, S. C. 1980. Notes on breeding and social behaviour of the White-winged Fairy-wren *Malurus leucopterus. Emu* 80: 157–61.

———. 1986. Breeding in three species of Fairy-wrens (*Malurus*): do helpers really help? *Emu* 86: 131–38.

Tidemann, S. C., Wilson, S. J. and Marples, T. G. 1988. Some results from a long-term bird-banding project in the Brindabella Range, A.C.T. *Corella* 12: 1–6.

Tinbergen, L. 1960. The natural control of insects in pine woods. I. Factors influencing the intensity of predation in song birds. *Arch. Neérl. Zool.* 13: 265–343.

Tomialojć, L., Walankiewicz, W. and Wesolowski, T. 1977. Methods and preliminary results of the bird census work in primeval forest of Bialowieza National Park. *Polish Ecological Studies* 3: 215–23.

———. 1984. Breeding bird community of a primeval temperate forest (Bialowieza National Park, Poland). *Acta Ornith.* 20: 241–310.

Turner, V. 1982. Marsupials as pollinators in Australia. In *Pollination and evolution,* pp. 55–66. *See* Keighery 1982.

Ueda, K. 1985. Juvenile female breeding of the Fan-tailed Warbler *Cisticola juncidis*: occurrence of two generations per year. *Ibis* 127: 111–16.

Underwood, L. A. and Stowe, T. J. 1984. Massive wreck of seabirds in eastern Britain, 1983. *Bird Study* 31: 79–88.

Van Rhijn, J. G. 1983. On the maintenance and origin of alternative strategies in the Ruff *Philomachus pugnax. Ibis* 125: 482–98.

Van Riper, C., Van Riper, S. G., Goff, M. L. and Laird, M. 1986. The epiizootiology and ecological significance of malaria in Hawaiian landbirds. *Ecology* 56: 327–44.

Vaughton, G. 1989. Pollination and seed-set of *Banksia spinulosa:* evidence for autogamy. *Aust. J. Bot.,* in press.

Vellenga, R. E. 1970. Behaviour of the male Satin Bower-bird at the bower. *Aust. Bird Bander* 8: 3–11.

———. 1980a. Distribution of bowers of the Satin Bowerbird at Leura, NSW, with notes on parental care, development and independence of the young. *Emu* 97–102.

———. 1980b. Moults of the Satin Bowerbird *Ptilonorhynchus violaceus. Emu* 80: 49–54.

Verheijen, J. A. J. 1964. Nesting seasons on Flores, Indonesia. *Ardea* 52: 194-201.

Verner, J. and Willson, M. F. 1966. The influences of habitat on mating systems of North American passerine birds. *Ecology* 47: 143–47.

Vestjens, W. J. M. 1975. Feeding behaviour of spoonbills at Lake Cowall, New South Wales. *Emu* 75: 132–36.

Vestjens, W. J. M. and Carrick, R. 1974. Food of the Black-backed Magpie *Gymnorhina tibicen* at Canberra. *Aust. Wildl. Res.* 1: 71–83.

von Haartman, L. 1971. Population dynamics. In *Avian Biology, Vol. 1:* eds D. S. Farner and J. R. King. New York: Academic Press.

Väisänen, R. A., Järvinen, O. and Rauhala, P. 1986. How are extensive human-caused habitat alterations expressed on the scale of local bird populations in boreal forests. *Ornis Scand.* 17: 282–92.

Waghorn, E. J. 1983. Population changes of the Australasian Gannet *Morus serrator* (Gray) at the Motu Karamarama Gannetry, Hauraki Gulf, New Zealand. *Emu* 82: 286–95.

Walters, J. R. 1984. The evolution of parental behaviour and clutch size in shorebirds. In *Shorebirds: Breeding behaviour and populations,* pp. 243–87, eds. J. Burger and B. L. Olla. New York: Plenum.

Warner, R. E. 1968. The role of introduced diseases in the extinction of the endemic Hawaiian avifauna. *Condor* 70: 101–20.

Weatherhead, P. J. 1986. How unusual are unusual events? *Am. Nat.* 128: 150–54.

Wegener, A. 1984. Animals killed by bushfires in coastal Victoria, February 1983. *Aust. Bird Watcher* 10: 248–50.

Weimerskirch, H., Jouventin, P., Mougin, J. L., Stahl, J. C. and Van Beveren, M. 1985. Banding recoveries and the dispersal of seabirds breeding in French Austral and Antarctic Territories. *Emu* 85: 22–33.

Weishu, H. and Purchase, D. 1987. Migration and banded waders between China and Australia. *Colonial Waterbirds* 10: 106–10.

Westerkov, K. E. 1971. Distribution and numbers of the Crested Grebe *Podiceps cristatus* in Canterbury. *Notornis* 18: 3–29.

Whelan, R. J. and Burbidge, A. H. 1980. Flowering phenology, seed set and bird pollination of five Western Australian *Banksia* species. *Aust. J. Ecol.* 5: 1–7.

White, F. W. G. 1986. Summer-winter movements of Olive Whistlers *Pachycephala olivacea* in the Snowy Mountains. *Corella* 10: 125–26.

White, G. 1789. *The Natural History and Antiquities of Selborne, in the County of Southampton.* London: Benjamin White and Son. (Edition seen — 1947 Cresset, London).

Whitmore, M. J., Dow, D. D., Fisk, P. and Moffatt, J. D. 1983. An annotated list of the birds of Meandarra, Queensland. *Emu* 83: 19–27.

Wiens, J. A. 1977. On competition and variable environments. *Amer. Scientist* 65: 590–97.

———. 1983. Avian community ecology: an iconoclastic view. In *Perspectives in Ornithology,* pp. 355–402, eds. A. H. Brush and G. A. Clark. Cambridge Univ. Press.

Wiens, J. A. and Rotenberry, J. T. 1980. Patterns of morphology and ecology in grassland and shrubsteppe bird populations. *Ecol. Monogr.* 50: 287–308.

Wiley, R. H. 1978. The lek mating system of the Sage Grouse. *Scientific American* 238: 114–25.

Williams, C. K. 1979. Ecology of Australian chats (*Ephthianura* Gould): reproduction in aridity. *Aust. J. Zool.* 27: 213–29.

Williams, C. K. and Main, A. R. 1976. Ecology of Australian chats (*Ephthianura* Gould): Seasonal movements, metabolism and evaporative water loss. *Aust. J. Zool.* 24: 397–416.

Williams, M. A. J. 1975. Late Pleistocene tropical aridity synchronous in both hemispheres. *Nature* 253: 617–8.

Willis, E. O. 1984. Conservation, subdivision of reserves and the anti-dismemberment hypothesis. *Oikos* 42: 396–8.

Wilson, E. O. and Willis, E. O. 1975. Applied biogeography. In *Ecology and evolution of communities,* pp. 522–534. *See* Diamond 1975.

Wingham, E. J. 1984. Breeding biology of the Australasian Gannet *Morus serrator* (Gray) at Motu Karamarama, Hauraki Gulf, New Zealand. II. Breeding success and chick growth. *Emu* 84: 211–24.

Winterbottom, J. M. 1949. Mixed bird parties in the tropics, with special reference to northern Rhodesia. *Auk* 66: 258–63.

Woinarski, J. 1985a. Foliage gleaners of the treetops, the pardalotes. In *Birds of eucalypt forests and woodlands: ecology, conservation, management,* pp. 165–75. *See* Bell 1985b.

———. 1985b. Breeding biology and life history of small insectivorous birds in Australian forests: response to a stable environment? *Proc. Ecol. Soc. Aust.* 14: 159–68.

Woinarski, J. C. Z. and Bulman, C. 1985. Ecology and breeding biology of the Forty-spotted Pardalote and other pardalotes on North Bruny Island. *Emu* 85: 106–20.

Woinarski, J. C. Z. and Cullen, J. M. 1984. Distribution of invertebrates on foliage in forests of south-eastern Australia. *Aust. J. Ecol.* 9: 207–32.

Wolf, L. L. and Stiles, F. G. 1970. Evolution of pair co-operation in a tropical hummingbird. *Evolution* 24: 759.

Wolf, L. L., Stiles, F. G. and Hainsworth, F. R. 1976. Ecological organisation of a tropical highland hummingbird community. *J. Anim. Ecol.* 45: 349–79.

Woolfenden, G. E. and Fitzpatrick, F. W. 1984. *The Florida Scrub Jay.* New Jersey: Princeton Univ. Press.

Wooller, R. D. 1984. Bill shape and size in honeyeaters and other small insectivorous birds in Western Australia. *Aust. J. Zool.* 32: 657–61.

Wykes, B. J. 1982. Resource partitioning and the role of competition in structuring *Lichenostomus* honeyeater (and *Manorina melanophrys*) communities in southern Victoria. Ph.D. thesis. Monash Univ. Melbourne.

———. 1985. The Helmeted Honeyeater and related honeyeaters of Victorian woodlands. In *Birds of eucalypt forests and woodlands: ecology, conservation, management,* pp. 205–17. *See* Bell 1985b.

Wylie, F. R. and Landsberg, J. 1987. The impact of tree decline on remnant woodlots on farms. In *Nature Conservation: the role of remnants of native vegetation,* pp. 331–2. *See* Saunders *et al.* 1987.

Wyndham, E. 1978. Birds of the Milparinka district and Cooper Creek basin. *Emu* 78: 179–88.

———. 1982. Movements and breeding seasons of the Budgerigar. *Emu* 82: 276–82.

———. 1986. Length of birds' breeding seasons. *Am. Nat.* 128: 155–64.

Wyndham, E. and Cannon, C. E. 1985. Parrots of eastern Australian forests and woodlands: the genera *Platycercus* and *Trichoglossus.* In *Birds of eucalypt forests and woodlands: ecology, conservation, management,* pp. 141–50. *See* Bell 1985b.

Wynne-Edwards, V. C. 1962. *Animal dispersion in relation to social behaviour.* Edinburgh: Oliver and Boyd.

———. 1986. *Evolution through group selection.* Oxford: Blackwell.

Yom-Tov, Y. 1987. The reproductive rates of Australian passerines. *Aust. Wildl. Res.* 14: 319–30.

Zahavi, A. 1975. Mate selection – a selection for a handicap. *J. Theor. Biol.* 53: 205–14.

Zamora, R. and Camacho, I. 1984. Evolución estacional de la communidad de aves en un encinar de Sierra Nevada. *Doñana Acta Vertebrata* 11: 25–43.

Index to Ecology of Birds

H

L

M